AGRICULTURAL SCIENCE

Gene Sequencing and Mapping

AGRICULTURAL SCIENCE

Gene Sequencing and Mapping

Editor

Ramniwas Sharma

BIOTECH

2006

Biotech Books

Delhi - 110 035

ISBN 81-7622-159-7

Published by : **BIOTECH BOOKS**
1123/74, Tri Nagar,
DELHI – 110 035
Phone: 27382765
e-mail: biotechbooks@yahoo.co.in

Showroom : 4762-63/23, Ansari Road, Darya Ganj,
NEW DELHI - 110 002
Phone: 23245578, 23244987

Printed at : **Chawla Offset Printers**
New Delhi – 110 052

PREFACE

Agriculture is the process of producing food, feed, fiber and other desired products by the cultivation of certain plants and the raising of domesticated animals. The practice of agriculture is also known as farming, while scientists, inventors and others devoted to improving farming methods and implements are also said to be engaged in agriculture

Agricultural chemistry deals with other vital farming concerns, such as the application of fertilizer, insecticides, and fungicides, soil makeup, analysis of agricultural products, and nutritional needs of farm animals. Modern gene sequencing, whether classical or through genetic engineering, comes with issues of concern, particularly with regard to food crops. The question of whether sequencing can have a negative effect on nutritional value is central in this respect. Although relatively little direct research in this area has been done, there are scientific indications that, by favoring certain aspects of a plant's development, other aspects may be retarded.

The emphasis may shift from gene mapping and genome analysis to the analysis of gene function and regulation, determination of genetic disease and somatic gene therapy. The development of novel data handling technologies may also be pursued. The opportunities for various genome projects have been discussed on the basis of advances in DNA sequencing technologies.

Editor

CONTENTS

CHAPTER 1

GENE CHARACTERISATION

'Characterisation' is the description of a character or quality of an individual. Characterisation of genetic resources refers to the process by which accessions are identified or differentiated. This identification may, in broad terms, refer to any difference in the appearance or make-up of an accession. In the agreed terminology of gene banks and germplasm management, the term 'characterisation' stands for the description of characters that are usually highly heritable, easily seen by the eye and equally expressed in all environments.

In genetic terms, characterisation refers to the detection of variation as a result of differences in either DNA sequences or specific genes or modifying factors. Standard characterisation and evaluation of accessions may be routinely carried out by using different methods, including traditional practices such as the use of descriptor lists of morphological characters. They may also involve evaluation of agronomic performance under various environmental conditions. In contrast, genetic characterisation refers to the description of attributes that follow a Mendelian inheritance or that involve specific DNA sequences.

The application of biochemical assays such as those that detect differences between isozymes or protein profiles, the application of molecular markers and the identification of particular sequences through diverse genomic approaches all qualify as genetic characterisation methods. Because of its nature, genetic characterisation clearly offers an enhanced power for detecting diversity (including genotypes and genes) that exceeds that of traditional methods. Likewise, genetic characterisation with molecular technologies offers greater power of detection than do phenotypic methods (e.g. isozymes).

This is because molecular methods reveal differences in genotypes, that is, in the ultimate level of variation embodied by the DNA sequences of an individual and uninfluenced by environment. In contrast, differences revealed by phenotypic approaches are at the level of gene expression (proteins).

MOLECULAR CHARACTERISATION

Information about the genetic make-up of accessions helps decision making for conservation activities, which range from collecting and managing through identifying genes to adding value to genetic resources. Well-informed sampling strategies for germplasm material destined for ex situ conservation and designation of priority sites for in situ conservation are both crucial for successful conservation efforts. In turn, defining strategies is dependent on knowledge of location, distribution and extent of genetic diversity.

Molecular characterisation, by itself or in conjunction with other data, provides reliable information for assessing, among other factors, the amount of genetic diversity, the structure of diversity in samples and populations, rates of genetic divergence among populations and the distribution of diversity in populations found in different locations.

A recent study on the genetic diversity of cultivated Capsicum species in Guatemalan home gardens compared the diversity present in an array of home gardens in the Department of Alta Verapaz with a countrywide representative sample of 40 accessions conserved ex situ in the national collection. The results showed that home gardens of Alta Verapaz (H = 0.251) contained as much diversity as the entire national ex situ collection (H = 0.281). These results thus suggest that,

— home gardens are indeed an extremely important resource for in situ conservation of Capsicum germplasm in Guatemala, and as such they should not be neglected;

— if further collecting activities were to be undertaken, special emphasis should be given to collecting in Alta Verapaz; and

— additional collecting in Alta Verapaz alone could disclose novel genetic diversity that is absent from the national collection.

Conservation of clonally propagated crops demands more complex and expensive procedures. If these crops are maintained on-farm, their existence is endangered by several factors, one of which being the introduction of alternative improved varieties. Conservation efforts need then to be based on solid knowledge of clonal diversity. This was the case for Abyssinian banana or ensete from Ethiopia, which was analysed with AFLP markers.

The results led to a reduced number of clones for conservation and indicated the existence of a common practice of exchange of local types between regions, which, in its turn, emphasised the need to collect further in different farming systems. Molecular characterisation helps determine the breeding behaviour of species, individual reproductive success and the existence of gene flow, that is, the movement of alleles within and between populations of the same or related species, and its consequences. Molecular data improve or even allow the elucidation of phylogeny, and provide the basic knowledge for understanding taxonomy, domestication and evolution.

As a result, information from molecular markers or DNA sequences offers a good basis for better conservation approaches. Management of germplasm established in a collection comprises several activities. Usually, such activities seek to ensure the identity of the individually stored and maintained samples, to ensure the safeguarding of genetic integrity and genetic diversity and to have the material available for distribution to users. These tasks are primarily a responsibility of gene bank managers and curators, and involve the control of accessions on arrival at the facilities, as well as their continuous safeguarding for the future through regeneration and multiplication.

For all these routine activities, information about the genetic constitution of samples or accessions is critical and provides possibly the most important means of measuring the quality of the work being performed. Bulk seed of wheat accessions to test their genetic integrity after 24 cycles of regeneration and after more than 50 years of storage at room temperature in a gene bank. They found neither contamination nor incorrect manipulation effects such as mechanical mixtures, but did identify one case of genetic drift in one accession. The fact that IPK-Gatersleben gene bank splits its germplasm samples into either almost or completely pure lines, i.e. accessions, is expected to have contributed to this very positive finding.

However, in the same gene bank, (1) a significant number of alleles present in the original sample was lacking in the newly regenerated material, and (2) new alleles in the new material were not present in the first regeneration sample. Thus, the use of molecular markers can quickly help check whether changes in alleles or allele frequencies are taking place. Molecular information has been used to weigh the need for decreasing the size of germplasm collections, which otherwise would add costs to the long-term conservation of germplasm.

For instance, Dean et al. used microsatellite markers to analyse the genetic diversity and structure of 19 sorghum accessions known as 'Orange' in the USDA's national sorghum collection. They found two redundant groups (involving five entries) among the 19 accessions evaluated. They also found that much of the total genetic variation was partitioned among accessions. Selected molecular technologies render cost-effective and comprehensive genotypic profiles of accessions ('fingerprints') that may be used to establish the identity of the material under study. Simultaneously, these technologies can detect contaminants, as well as the presence of redundant materials (or 'duplicates').

Moreover, molecular data provide the baseline for monitoring natural changes in the genetic structure of the accession, or those occurring as a result of human intervention (e.g. seed regeneration or sampling for replanting in the field). Whatever the case, analysis of molecular information allows the design of strategies for either purging the consequences of inappropriate procedures or amending them to prevent future inconveniences. A small number of potential duplicates were identified in a core collection of cassava when isozyme and AFLP profiles were compared.

The core collection had been assembled with information from traditional markers, which proved to be highly effective for selecting unique genotypes. Molecular data were used for efficiently verifying the previous work on the collection and ensure minimum repetition. The taro core collection for the Pacific region was treated in a similar manner. Thus, gene bank managers can easily realise the potential value of using molecular methods to support and possibly modify or improve a gene bank's operations.

A special and increasingly important role of genetic characterisation is that of identifying useful genes in germplasm, that is, of maximising conservation efforts. Because the major justification for the existence of germplasm collections is use of the conserved accessions, it is important to identify those valuable genes that can help develop varieties that will be able to meet the challenges of current and future agriculture. Characterisation has benefited from several approaches resulting from advances in molecular genetics such as genetic and QTL mapping, and gene tagging.

The acknowledgement of the value of wild relatives, in which modern techniques have discovered useful variation that could contribute to varietal improvement. Knowledge of molecular information in major crops and species and of the synteny of genomes, especially conservation of gene order, has also opened up prospects for identifying important genes or variants in other crop types, particularly those that receive little attention from formal research.

Target Specific Genomic

Most marker technologies target genomic regions, which are selectively neutral; some technologies, however, target specific genes. The neutrality of markers is suitable for most uses in germplasm conservation and management. However, when the interest of conservation lies specifically in the diversity of traits of agronomic importance, some questions remain about the markers' representativeness. In such a case, those markers able to detect functional diversity are more suitable for characterising germplasm collections.

Germplasm in collections can undergo molecular characterisation that is structural, meaning 'based on the building blocks of the DNA sequence', and functional, that is, based on the identification of genes and their functions. Such characterisation permits access to the raw materials-the genes-for nearly all the objectives of today's and tomorrow's breeding programmes. The information gathered from structural characterisation not only provides increased clarity on existing genetic diversity and its organisation in individuals, but it determines sample and population organisation that ultimately may form the basis for functional characterisation.

The increasing number of sequencing projects has resulted in an increased opportunity to produce expressed sequence tags (or ESTs) to which gene functions may be assigned. Moreover, such projects are making possible the compilation of an enormous amount of sequence data that can be used to develop markers linked to specific genes, which, in turn, may help identify novel functional variation.

In addition, the development of novel technologies continues. This usually means decreased costs-a very significant point for their application in the tasks of conserving genetic resources, which tend to involve large numbers of samples and to have difficulties in sourcing needed funds. Other improvements involve increasing the throughput-both in number of markers analysed and in number of samples-and simplifying technologies. New developments are also taking place in designing better approaches to access new and useful genetic variation in collections, namely, allele mining and association genetics.

Allele mining focuses on the detection of allelic variation in important genes and/or traits within a germplasm collection. If the targeted DNA (either a gene of known function or a given sequence) is known, then the allelic variation (usually point mutations) in a collection can be identified, using methods developed for the purpose. Association studies of artificial progenies are an alternative to segregation analysis for identifying useful genes by correlation of molecular markers and a specific phenotype.

Association studies can be performed on a germplasm collection and also on other materials, as long as significant linkage disequilibrium (LD) exists, for example, breeding materials. It may be especially useful for those crops where appropriate populations for genetic analysis cannot be obtained or their production is too time-consuming. It is also useful for those crops for which sequence information does not exist and is unlikely to be available soon.

The importance of the variation captured in genetic resources in allowing evolution and/or facilitating plant breeding has been long recognised. However, appreciating the variation held in collections is not sufficient. Conservation of genetic resources needs to go hand in hand with enhanced use of the conserved material. Identifying and making available the allelic variation that makes up the genotype and phenotype provides the groundwork on which genetic resources can be used in, for example, plant breeding.

The number of accessions held collectively by all CGIAR gene banks is estimated as being almost 600 000. Together with the collections established by national programmes worldwide, this number reaches almost 6 million. Without doubt, these genetic resources collections, together with uncollected germplasm and that held in situ and on-farm, harbour abundant quantities of hidden allelic variants. The challenge is to unravel the mysteries of this variation so it can be used for the benefit of humankind.

Gene banks hold large numbers of accessions, particularly of staple crops. Modern improvements in equipment and procedures allow considerable sample throughput. This can be costly. However, the more a technology develops, the lower its costs will be per data point and per sample. Nevertheless, the higher the throughput used, the higher the number of data points obtained. This requires adequate equipment-for handling and storing-and expertise-for handling and analysing-to draw adequate results from the investment.

One possible avenue for ensuring broader benefits from molecular characterisation is the establishment of international collaboration for particular crops. Although, currently, equipment and expertise cannot be readily available worldwide, characterisation networks are possible. In addition to carrying out the laboratory work, such networks would also facilitate access to information, thus fostering closer links between curators, breeders and molecular scientists. At the same time, those countries with little expertise or equipment can make steady progress in both areas and, hence, can make better use of the genetic resources that they hold.

More and more, technologies have increased throughputs, which generally means the generation of progressively larger amounts of data. Such data should not languish unused. If gene banks equip themselves with the latest technologies, then they should be able to translate such data into scientific knowledge. To do so, they need not only laboratory technical expertise, but also need bioinformatics staff. This means that through their molecular work, gene banks may keep not only live plant materials but also DNA and data.

Hence, the banks may develop appropriate new roles as providers of genetic resources and their accompanying data in an array of forms. Such broadening of the gene banks' roles implies that their clientele will also expand from plant breeders to include molecular geneticists, molecular biologists and even bioinformaticists. The expanded range of roles may even lead to including activities related to phenotyping, a type of characterisation beyond the traditional description of morphology and general field performance. Phenotyping is very much linked to the usefulness of good molecular characterisation as, together, they form the basis of progress in modern genomics.

CHAPTER 2

GENETIC RESOURCES AND GENE-BASED INVENTIONS

Agriculture is an example of a 'complex technology system' of varied techno-economic institutions that exhibits subtle interdependencies and component compatibility. Inputs build the environment and provide the general conditions in which seeds perform; and it is necessary for seeds and inputs to be compatible for an effective exploitation of the latent economies in the technological package. However, seeds occupy a particular primacy in that they establish the upper limit to the productivity of all other agricultural inputs. The primacy of seeds is on account of the dual properties embedded in the 'seed'. A packet of seeds is composed of two distinct and separable properties: (a) genetic information, i.e. software, which is the result of breeding programmes, and (b) physical properties, i.e. diskette-like features that are determined by seed production process.

In reality is the genetic software, i.e. the varietal characteristics, that is of crucial importance in substantially determining the productivity limits of agricultural inputs. For example, while the introduction of inorganic chemicals relax the constraints of renewing soil fertility after each cycle of cultivation, transformations at the varietal level are necessary to enable crops to withstand and effectively utilise such applications. Economic returns associated with the use of these inputs depend on the adoption of varieties that are bred to respond to the application of these inputs. A similar argument can be made for the case of mechanisation, where effective mechanisation of farm practices requires specific transformations in the biology (e.g. uniform maturing) and architecture (e.g. strong stalks) of the plant.

Consequently, mech anised harvesting of certain crops (e.g. tomatoes) had to await a breeding achievement (e.g. hard exteriors). Yet, neither specific input packages, nor particular types of varieties (e.g. dwarfs) can independently proliferate as the dominant mode of agriculture. Rather, it is a technology package consisting of specific varieties that are compatible with defined inputs and agronomic practices. However, the malleability of plants, on account of changes achieved in its genetic software give it (i.e. the 'seed') a unique position of being a platform for the techno-economic transformation of agriculture. The characterisation of 'seeds' as platforms for technical change highlights another aspect of the component-interdependencies of the technology system.

Changes achieved in genetic software that tie together variety and input enable the creation of markets for industrially produced agricultural inputs. Because of the systems nature of agricultural technology, these different markets are increasingly interlinked. In the era of biotechnology, evidence of the interlinking of agricultural product markets via transformation achieved at the level of the genetic software is more transparent. Take for example the breeding focus on herbicide-tolerance: adoption of a particular variety ties the farmer to an input package consisting of brand-name herbicides. Not surprisingly, herbicide tolerance is the dominant trait being tested across all genetically modified crops: the proportion of transgenic crops with this single trait increased from 23% to 54% of the total global area under transgenic crops between 1996-97.

CHANGING ROLES OF THE PUBLIC AND PRIVATE SECTORS

Producing plant varieties for use by farmers require a number of different activities, each with its distinct techno-economic resource requirements. Plant breeding is but one of the many steps to be accomplished. The following schema appears well accepted within the literature:

Varietal development: The development of genetic software (i.e. new varieties) involves assembling diverse genetic material, crossing the genetic material to generate variation, selecting recombinations from the variation, and stabilising the preferred characteristics as a new variety. The specific method of selecting preferred characteristics differs according to the mode of propagation of the species (e.g. cross or self-pollinating) and the type of breeding method (e.g. hybrid or pure line). The varieties are tested and evaluated for specific characteristics and than released as breeders' seed.

Seed production and multiplication: It is here that the diskette (i.e. the seed) carrying the selected genetic software (i.e. varietal characteristics) is produced through the sequential multiplication of breeder seeds. The preferred method of multiplication depends on the mode of propagation of the crop, while the number of multiplications depends on the regulatory system. It is quite common to have a three-stage process that sequentially produces 'foundation seed', 'certified seed' and 'commercial seed'. It is important to maintain the distinguishing characteristics of the variety, so that the farmer plants a diskette (i.e. seed) that corresponds to the advertised claims of the software (i.e. variety).

Seed processing and storage: Given that seeds are living organism, a number of precautionary steps need to be undertaken to maintain the viability of the seed over a period of time between production and eventual utilisation. These precautions involve steps like drying, cleaning, and chemical treatment.

Seed marketing and distribution: This involves a number of activities, such as transportation, promotion, field demonstrations, advertising, etc., to ensure that the right amount of seed of the appropriate variety reaches the farmer at the correct time. Equally crucial is the requirement of informing farmers of the characteristics and agronomic performance of available varieties so as to enable appropriate and informed decision-making by farmers.

In addition to the activities listed above are functions like germplasm collection and documentation and fundamental research. These wider activities provide the environment within which plant breeding can take place and are invariably performed by the public sector.

The literature on seed systems makes policy prescriptions on the appropriate role for the public and private sector in terms of the activities identified above. The analysis is based on mapping seed system development across countries, which has led to the identification of three broad stages. In the early stages of seed system development farmers tend to be the primary and dominant source for varietal improvement. The transitory middle stages are characterised by the dominating presence of public sector breeding activities and mixed presence of private-public agencies in downstream seed system activities. At this stage, as far as varietal development is concerned, private sector investments tend to focus on highvalue/ low-volume commercial crops and hybrids.

The final stage of a mature seed system is distinguished by the presence of private sector companies in varietal development. Underlying this analysis are stereotypical characterisations of the private and the public sectors. Thus, the public sector is presented as being 'inefficient or ineffective'. In contrast, the "private sector is normally flexible, and responsive to changing requirements". However, there is little disagreement that, following the adoption of policies of liberalisation, the public sector and parastatal organisations are under significant budgetary pressure, which undermines its capacity to deliver varietal improvements.

Consequently, the literature recommends the gradual withdrawal of the public sector from 'near-market' activities and from activities that allow high levels of appropriation. These policy prescriptions are reflective of the continuous struggle to demarcate the private and the public domain whilst balancing the interests of different social groups engaged in agriculture. No doubt, often there are strong views expressed on the policies that inhibit the development of a commercial seed industry, as expressed in the following quote, "The disincentives and uncertainty caused by seed distribution programs, the inefficiencies and privileges of public seed producers, and the existence of restrictive regulatory regimes all conspire to inhibit the development of a commercial seed sector in Africa".

Among the policies recommended for the promotion of a commercial seed industry is the introduction of intellectual property protection for plant varieties. It is suggested that the availability of IPRs will enhance the scope of appropriability for commercial breeders, thus improving the incentives for investments in plant breeding, which will eventually lead to the development of better varieties. However, this policy prescription misses two crucial factors that underpin the development of a commercial seed industry. First, on the demand side, it is necessary for the seed market to exhibit adequate and sustained purchasing capacity.

At a global level, the commercial seed industry accounts for approximately a-third of the global market and is valued at US$30bn. The balance is divided equally between the public sector and a locally/farmer-supplied system. Disaggregating this to regional and local levels reveals the marginal position of commercial seed providers. For example, only about 7% of wheat seed and 13% of rice seed in India is sourced from the formal (public and/or private) sector, and in many parts of Africa and Asia it is estimated that over 80% of total farmers' seed requirement are met from outside the formal sector. In most instances, farmers source seed from a variety of outlets that include neighbours, farm-saved seed/grain, and local seed multipliers.

In addition, most farmers do not regularly purchase seeds for each successive planting. This is not peculiar to farmers in developing countries, since wheat farmers in both industrialised and developing countries exhibit a varietal turnover rate that ranges between 5-10 years. Part of the relatively low and irregular demand for fresh seeds in developing countries is a reflection on the inappropriateness of breeding strategies within the formal sector and the pricing policies of commercial seed producers. Second, the viability of a commercial seed industry is significantly contingent on the preexisting research base and enabling scientific environment.

The technology studies literature has amply demonstrated that it is public science which substantially provides the leads and opportunities for the private sector to pick up and pursue. In the case of plant breeding there are a number of activities that are mainly conducted by the public sector, such as germplasm documentation and maintenance, development of breeding lines, training of scientists and breeders. Unfortunately most accounts of seed system do not pay adequate attention to this factor. Thus, even in a 'mature' seed system such as the US, where PBRs have been available since 1970, the dependence on the public sector is significant and apparent in the parentage of privately bred varieties: 50% of the wheat and soybean varieties, 90% of barley and dry bean varieties and 95% of rice varieties consist of publicly bred varieties; in 1979, 72% of hybrid corn varieties had at least one parent of public sector origin. These parental lines have invariably been accessed at nominal costs of about US$5-20.

To sum up, recognition of the properties of the seed is of primary importance for a policy analysis of intellectual property rights in plant genetic resources. In this respect, it is important to recognise the malleability of plants, on account of changes achieved in at the level of its genetic software, which allow the seed to occupy a position of the platform for the techno-economic transformation of agriculture. The discussion also highlights the general policy prescription for increasing privatisation of different components of the seed system. These recommendations require urgent re-evaluation in light of (a) factors defining the demand conditions for seeds and (b) the supporting role of public sector breedingrelated activities.

ECONOMIC IMPACT OF PLANT VARIETY PROTECTION

The cornerstone and fundamental social logic of providing IPP is to promote socioeconomic progress through granting temporary monopoly rights to inventors, which allows them to appropriate a rent from the use/exploitation of their inventions. The protection provided is often weak because there can be many viable solutions to a technical problem, so other firms can "invent around" a given patented solution. For smaller and especially less-developed countries, patent holders' power to block use of their inventions by others poses a special problem.

Multinational corporations commonly patent their most important inventions in dozens of national jurisdictions. The quest for scale economies leads them to produce in one or a few preferred locations ... this typically means that high prices will be paid for imported patented products, while opportunities to build a home industry using first-line technology are restricted. It is to issues like those referred to in the above quote that attention needs to be devoted - the evidence of the actual working of IPRs in agriculture. Again, a focus beyond the available 'checks and balances' in the letter of the law, is not novel.

Impact of PBRs in Developed Countries

It is often claimed, that at a minimum, IPRs operate as "anticipated" and result in increased investments. This presumption about the impact of IPRs raises two questions. First, is this increased breeding activity and investment equally distributed across all company? Evidence from the US clearly demonstrates that older companies, i.e. companies that pre-existed the introduction of PBRs, are more R&D-intensive, have a larger scale of operation and breed across a broader crop portfolio. It appears that older firms have an advantage reflecting their accumulated scientific and technological competencies. However, the evidence also concludes that this advantage is exacerbated by the acquisition of IPRs.

Second, is the increased investment and breeding activity evenly distributed across crops? The only evidence on the distribution of R&D across crops is from the US, where it is selfevident that a few crops (i.e. wheat and soybean) experienced increases in private investment. Unfortunately, there is nothing to suggest that the increases were primarily or entirely on account of the availability of PBRs. With respect to soybean, a range of factors influenced the investment behaviour of private breeders, such as (a) historical trend of increasing acreage under soybean cultivation, (b) fragility of the soybean seed, which disallows farm-seed saving, (c) expanding downstream food processing industry based on soybean, and (d) increasing international trade. The increase in wheat breeding activity - the other crop to account for sizeable private investments - is explained by the concentrated focus on developing hybrids.

Given that PBRs are provided for new varieties, economists have focussed on the release of new varieties as an indicator of the impact of IPRs. A general claim in the literature is that "the availability of PBRs has increased the number of private sector breeders, as well as the number of varieties released and planted". The claim is (partially) supported by evidence from the US and time series data of UK wheat varieties. Interestingly, even critics accept that the number of varieties released have increased after introduction of PBRs. Recent evidence for the US suggests that the number of varieties released in the 1990s is higher than that in the 1980s.

However, key methodological questions remain in terms of whether the availability of protection caused the increase in varietal release and whether this is an economic good. Three questions remain before a conclusive statement on the impact of PBRs on varietal release rates can be accepted as an economic good.

First, has there been a change in the historical rates of varietal release in the pre- and post-PBRs period? Naturally, a historical examination of changes in the varietal release rates will also take account of changes in other factors that influence breeding activity. No such study has been conducted. The only effort at partially analysing historical rates of varietal release for select UK crops across 1930-90 reports mixed evidence: while rates of varietal release, following introduction of PBRs, increased in apple, the change in French beans was marginal and it was negligible in strawberry.

Second, what are the agro-economic qualities of the new varieties? This is important, since an increase in the availability of varieties is not an economic good in itself, as it might be that the increase in varieties is on account of cosmetic breedingand/or strategies of planned obsolescence. Further, we must keep in mind that there is no merit test in the process leading to a PBRs grant. It is suggested that "protected varieties will be shown to be good performers". The claim is based on reports of field trials that aim at disaggregating yield increases amongst different components, which conclude that the 'breeding effort' component is substantial and has increased over time. While there is little doubt that new and contemporary varieties are more productive, it is difficult to disentangle the interactions between the adoption of new varieties and the use of input mixes and crop husbandry practices, since the varieties are themselves selected to respond to the inputs and husbandry practices.

Third, are changes in varietal release rates part of wider appropriation strategies? At issue here are possible interactions between the process of technical change and the appropriation strategies, which manifest in the entry-exit dynamics of varieties. To explain, firms might adopt strategies of planned obsolescence as a means of maintaining market shares, which result in faster rates of varietal turnover and higher varietal release rates. This strategy is underpinned by breeding efforts that focus on incremental productivity improvements (e.g. yield increases) and reduced durability (e.g. narrow disease resistance profiles). Both actively reduce the useful economic life of a variety, and thus warrant adoption of the new vintage by the farmer. Empirical evidence from wheat in the UK reveals that the average age of varieties fell from over 12 years to about 6 years 1960-95.

The three 'questions' place evidence of changes in varietal release rates in a wider context. To sum up, while it might be the case that more varieties are annually released under a regime of PBRs, it is not automatically the case that this is an economic good. In fact, empirical evidence of strategies of cosmetic breeding and planned obsolescence cast doubt on this claim. Finally, there is little doubt that new and contemporary varieties are increasingly productive. However, efforts to claim that this is substantially a result of the 'breeding effort' and not of the wider package (inputs and agronomic practices) are not convincing.

Market concentration

Concerns about market concentration are quite natural in debates on PBRs. But, first some evidence. The distribution of PBRs in wheat in the UK is highly concentrated and the degree of concentration has increased with time: the top five grant holders accounted for 69% of the grants in 1965-74 which increased to 79% in 1986-95. In the case of the US no comparable time series data exists. Evidence of the grants issued between 1971-82 provides the following conclusions:

In marigold, alfalfa and oats the top 2 grant-holders accounted for more than 70% of the grants. In pea, bean, lettuce, watermelon, and barley, the top 3 grantees accounted for more than 50% of the grants, whereas in tobacco they held 100% of the grants. The top four grant-holders accounted for between 82-91% of the grants in onion, rice, tomato and cauliflower. These levels of concentration are high, and it is suggested that seed firms that were established well before the passage of the Plant Variety Protection Act have benefited. More importantly, the adverse crop-level distribution of grants corresponds to seed market concentration.

Commenting on the US data, Lesser argues that in the long run, competition pressures will eliminate the negative effects of market concentration. Yet, recent data on the North American seed market indicates the contrary: Hybrid corn: the top five companies account for 69% of the market, with the market leader, Pioneer Hi-Bred, alone controlling 42%.

Soybean: the top five companies account for 51% of the market, with the market leaders, Pioneer Hi-Bred and Monsanto, each controlling 19%. Cotton seed: the market leader, Monsanto, alone controls 84% of the market, on account of its purchase of Delta and Pine Land. Similar evidence is available in the case of wheat in the UK where the market leader, Plant Breeding International, has consistently controlled 60-80% of the market through the period 1977-95. Data of the global seed market is notoriously unreliable and difficult to collect. One effort at collating information from company annual reports indicates that, in 1998, US$30bn of the US$50bn global seed market was accounted for by commercial seed sales - of which, 31% was accounted for by the top 23 companies, with the top 3 accounting for 13%. Another source suggests that the top ten companies account for 30% of global commercial seed sales.

It is the exercise of market power that is of concern, which can be discerned from movements in seed prices. The evidence is not encouraging: US, expenditures on seeds increased from US$519mn to US$1515 between 1960 and 1980. Three-fourths of the increase was on account of seed price rise and the balance on account of increased seed purchases. In the US, over the 1972-92 period, real seed prices increased faster than yields. In the UK, with the removal of controls on royalty rates, farmers are paying more for seeds. The evidence noted above clearly reveals that a high level of market consolidation has developed in the seed industry. More importantly, the evidence of movement in seed prices is suggestive of an undue exercise of this market power.

The introduction of PBRs has differential results across crops. More over, the increasing presence of the private sector in plant breeding is accompanied by appropriation strategies that might have deleterious distributional and allocative impact. For example, increase in the varietal release rate is accompanied by a shortening life-span of varieties. Equally important is the evidence of correspondence between concentration in PBRs grants issued and seed market shares, which manifests has led to increases in seed prices.

A wide variety of commentators accept that there is 'little evidence' or 'mixed and inconclusive evidence' about the direct benefits of introducing IPRs in plant varieties in developing countries. Others agree that the costs of stronger intellectual property protection are easily foreseeable, hence the need for identifying and assessing the wider social welfare implications and possible benefits. Yet, the enactment of IPRs in developing countries is recommended, as in the following statement; If the LDCs do not experiment with strengthened IPRs as a means of bringing more private innovative activity into the effort to capitalise on this new technology, the productivity gap will surely widen.

In contrast, a more cautious approach is also present in the literature. Thus, Barton suggests that it is "doubtful" whether developing countries should enact PBRs because the "tradeoffs are quite different". In this respect, contrast the estimate that by 2000, 75% of US agricultural production would be accounted for by 50,000 large farms, with the estimate that 70% of India's population is dependent on agriculture. Consequently, a detailed empirical analysis of the effects of adopting PBRs is considered a necessary prerequisite before policy formulation/ implementation. In this respect, the key questions for policy analysts in developing countries prepared by IPGRI are useful.

At issue here is whether the introduction of IPRs in developing countries will lead to increasing plant breeding activity that meets the agro-economic needs of the country. As with the differential focus on select crops, available evidence of private breeding in developing countries indicates a limited breeding effort that concentrates on few crops that tend to be hybrids and developed for exclusive agro-ecological zones. Evidence in this respect is available in an unpublished USAID study on Argentina and Chile. The private sector invested only in wheat and soybean. The only comprehensive study of the impact of PBRs in a developing country setting analyses evidence for five Latin American countries.

Among the findings, the following is pertinent in terms of the breeding effort: de facto division of labour has evolved with MNCs predominantly operating in hybrid crops and the public sector (and domestic breeding companies) almost entirely focussing on open-pollinated varieties. The MNC-funded breeder lobby group advocated and secured the inclusion of protection of inbred parental line within Argentina's PVP legislation, and consequently, the number of protected inbreds increased to 112 in two years. Lesser, quoting CGIAR personnel, accepts that agricultural research for developing countries is dedicated to few crops and tends to be suitable for 'post-Green Revolution' areas. Unfortunately, this type of research focus does not relate to low-input areas where the diversity and variability of growing conditions cannot be normalised through the purchase of inputs, thus, places a greater premium on breeding for local adaptation.

As such, the likelihood of private breeding meeting the agro-economic needs of the vast majority of farmers is unlikely. This view is reiterated in the following quote: The assumption that new varieties are always 'on the shelf', ready to be delivered to farmers, has been disproved by the disappointing experience of many extension programmes meant to deliver supposedly superior varieties that have not been sufficiently tested under farmer conditions. The demand for seeds of new varieties, especially in the farming conditions that Chambers describes as 'complex, diverse, and risk-prone' will depend on the ability of modern plant breeding to adopt strategies that are better able to target these varied environments and to develop appropriate varieties.

Reflective of the above statement is the difference between the public and private sector in maize breeding in India (Rangnekar, forthcoming). Maize breeding has attracted sizeable private sector investments in India, such that it almost matches the public sector in varietal release rates. However, there are striking differences in the varieties which reveal the agro-economic areas being targeted:

Hybrids: the private sector exclusively breed hybrids, whereas only 23% of public sector varieties are hybrids Varietal Characteristics: private sector hybrids are suitable for full-season and irrigated areas, whereas public sector varieties are hard, flinty grain types that are suitable for small-scale farmers in vulnerable agro-ecological niches It may be suggested that private sector focus in biotechnology might be relevant. In this respect, rice is an illustrative example with over 160 patent grants issued between 1982 and 1997. The research programmes focus on hybrid rice and geneticallymodified rice, raising question whether predominant production problems confronting developing countries are being addressed.

Of interest is a Monsanto presentation at a recent IRRI conference, which emphasised the attraction of the seed market, net present values estimated at US$1-2bn. In addition, the interest in rice is motivated by the availability of "mechanisms to add value, mechanisms to capture value and mechanisms to preserve value [i.e. intellectual property rights]". Clearly, it is the opportunities thrown up by scientific/technological developments, the size of the seed market and appropriation possibilities through IPRs that explain the evidence. Yet, in the absence of patent protection in many Asian developing countries, PVPs could provide some incentive.

As far as global agri-biotech research is generally concerned, there is little focus on the needs of poor farmers. The main crops targeted are soybean, corn, tobacco, cotton and canola together account for over 95% of the acreage. Further, only two traits dominate: insect resistance using Bt and herbicide tolerance. However, herbicide tolerance appears as the preferred trait being pursued, with area under cultivation increasing from 23% in 1996 to 54% in 1997. To sum up, existing evidence of the focus of private sector plant breeding is not entirely promising.

The range of crops focussed on and the type of agro-ecological niches being targeted do not cater to the wider needs of the majority farming populations in developing countries. It is often suggested that this will change with the availability of IPRs; however, the trend within biotechnology is not promising. On the other hand, it must be reckoned that the vast majority of farmers in developing countries are not a market with purchasing power as a substantial segment of the market is provided by non-commercial transactions or local production. This raises a deeper 'chicken-and-egg' question: is it the absence of effective demand that forestalls the supply of suitable plant varieties; or is the lack of suitable varieties that stunts socio-economic growth in these areas?

Access to Varieties and Germplasm

Another benefit of the availability of IPRs in developing countries is that it will enable access to varieties bred outside the country, as foreign breeders would be more willing to provide access to their varieties when IPRs are available. This appears to be the case for an unpublished USAID study of Argentina and Chile, where access to US and New Zealand fruit varieties improved following the introduction of PBRs. It is similarly argued that Monsanto refused access to Bt-cotton in Brazil, despite extensive crop loss from infestation, because of absence of proprietary protection.

Consequently, the claim that the absence of intellectual property protection is a restriction on the access to germplasm. In this respect, the experience of Kenya, where a cut flower industry has flourished, makes interesting reading. No doubt, Kenya has benefited, to some extent, from the development of a cut-flower industry. However, whether this has improved food security remains a moot point. There is more to the question of access to breeding material. A five country study in Latin America concludes that access to germplasm was significantly restricted after the implementation of PBRs - a result that is consistent with evidence from the US studies reviewed earlier. The transformations that take place are more subtle than apparent. Introduction of PBRs is often followed by dual processes that (a) place pressure on public institutions to behave like 'semiprivate organisations' and (b) motivate closer collaborative research between the public and private sectors.

Public Plant Breeding

Public plant breeding remains a key component of the agriculture research systems in developing countries. Its importance goes beyond the obvious role of breeding varieties suitable for the diverse local needs and includes the role of providing the scientific background and technological opportunities for private sector developmental activities. However, public agricultural research will have to undergo a radical transformation in an era marked by the increasing presence of the private sector. This change is partly a consequence of a deepening resource crunch precipitated by the reluctance of international donors to finance agriculture-based activities.

Three closely interconnected issues confront public agriculture research in the future. First, the growth of private sector research activities has led to reconfiguring the social division of labour between public and private. Here it is crucial to remember that neither one of the two sectors can meet the wide and diverse needs of all users of agricultural research. This is also implies that the research activities conducted by the public and private sectors are not substitutes for each other. In a public welfare sense, the research programmes in the two sectors are non-substitutable. Moreover, as the public sector is increasingly expected to meet the needs of those segments of the agricultural population that lack purchasing power, the possibility of internal revenue generation (or cost recovery) for the public sector is limited. This should place ever greater pressure on international donors to maintain, if not increase, the real funding for public agricultural research.

Second, and closely following the above, there are questions about the emerging relationship between the public and the private sector. Concerns here are warranted by changes in the nature of interactions between the two. Initially fostered by a greater revenue orientation in the US following the Bayh-Dole Act that promoted the transfer of research results from the public sector to the private sector. There remains an uneasy and troubling alliance in these ventures, which are "part of the public research domain but [behave] with the profit motive of a regular firm". These transformations have been observed within developing countries and even within international agricultural research centres. This increased 'networking' between the public and the private sectors raise vital policy questions.

To be clear, the issue is not that 'networking' between public and private research centres is to be avoided. Not at all, in fact to the contrary - networking is an important and essential part of the organisation of research that allows the effective exploitation of knowledge spillovers and exploration of a wider technological domain. Moreover, in today's complementary research environment this might also be considered essential. Rather, issues of concern involve accountability of the public sector institution, transparency, dynamics of the transfer of proprietary research and control over dissemination of research results.

Third, there are considerations about the impact of proprietary control in research tools on the viability and functioning of public agricultural research.

TRIPs AGREEMENT AND PLANT INNOVATIONS

The TRIPs Agreement expresses three broad objectives in its Preambular statement: (a) establishing the minimum standards of protection in terms of the main instruments of intellectual property protection, (b) clarifying the general principles concerning the domestic procedures and remedies for enforcement of IPRs, (c) making disputes between WTO members regarding TRIPs obligations subject to the WTO's dispute settlement procedures.

Patentability of Plant Genetic Resources

At issue concerning the patentability of plant genetic resources are the criterions for granting patents. Any invention applying for patent protection under a TRIPs-compatible system must necessarily fulfil the 'normal tests of novelty, inventiveness and industrial applicability' as given in the Agreement (WTO nd,

'Overview of the Agreement'). These conditions exist "in one or another form" in practically every country which has a patent system. However, none of the concepts - invention, novelty, inventiveness and industrial applicability - are defined in the Agreement.

Neither is US law particularly helpful. Rather, US law, without providing a legal definition for invention, identifies the subject matter that may be invented as including any useful process, machine, manufacture, or composition of matter or any useful improvement thereof. Consequently, much attention is devoted to the procedural requirements of patenting as a means to draw the line between patentable and non-patentable subject matter.

Defining Plant Varieties

There is a vast literature on 'defining plant varieties' in terms of UPOV and the EPC. It is fair to say that the founding principles of UPOV reflect a pragmatic approach aimed at achieving a 'working definition' of plant variety that, while corresponding to the activities of the breeder, also provide a legal basis for protection. The resulted in an approach focussing on the identifiability of a plant variety, based on the fixity of distinguishing characteristics, as the means of defining the subject matter of protection. Consequently, the method of identifying plant varieties - the triple requirement of distinctness, uniformity and stability - becomes the legal concept of a plant variety. Inherently, this approach, which has been christened the 'snap-shot' requirement, conflicts with, among other things, the variability of plant genetic material.

The term sui generis, however, may offer a wider range of policy choices because it could, presumably, include any arrangement for plant varieties that offers recognition to innovators - with or without monetary benefit or monopoly control. The statement captures the diversity of views and possible implementation options and positions noted above. Conceivably, in the absence of any overt/covert coercion, there will be a level of disharmony. This in itself should not be alarming as the Agreement, despite efforts at establishing uniformity in core standards and procedures concerning IPRs, does not seek to achieve a global harmonisation of domestic IP law.

More importantly, "even if the exercise of these options [i.e. exclusion clauses and ambiguities in definitions] must remain consistent with the express requirements of the Agreement" a lack of harmony is inevitable. A number of commentators agree that available flexibility in the Agreement, in general and also with respect to the sui generis option, means that there might be a 'disharmonising effect'. In fact, WTO officials have acknowledged the 'disharmonising effect' as a fallout of the lack of a definition of 'effective' and the absence of a reference to any existing international Convention.

CHAPTER 3

INHERITANCE AND MOLECULAR MAPPING OF GENES

Wheat is the most important staple food of about two billion people. Worldwide, wheat provides nearly 55% of the carbohydrates and 20% of the food calories consumed globally. It exceeds in acreage and production every other grain crop and is therefore, the most important cereal grain crop of the world, which is cultivated over a wide range of climatic conditions and the understanding of genetics and genome organization using molecular markers is of great value for genetic and plant breeding purposes.

The grass family Poaceae (Gramineae) includes major crop plants such as wheat (Triticum aestivum L.), barley (Hordeum vulgare L.), oat (Avena sativa L.), rye (Secale cereale L.), maize (Zea mays L.) and rice (Oryza sativa L.). Triticeae is one of the tribes containing more than 15 genera and 300 species including wheat and barley.

Wheat belongs to the tribe Triticeae (= Hordeae) in the grass family Poaceae (Gramineae) in which the one to several flowered spikelets are sessile and alternate on opposite sides of the rachis forming a true spike. Wheats (Triticum) and ryes (Secale) together with Aegilops, Agropyron, Eremopyron and Haynalidia form the subtribe Triticineae.Linnaeus in 1753 first classified wheat. The chromosome number sets (genomes) for each commonly recognized type.

This was a turning point in Triticum classification. It separated wheat into three groups. Diploids had 14 (n=7), tetraploids had 28 (n=14) and the hexaploids had 42 (n=21) chromosomes. Bread wheat is Triticum aestivum. T. durum and T. compactum are the other major species. All three are products of natural hybridization among ancestrals no longer grown commercially.

Wheat is an edible grain, one of the oldest and most important of the cereal crops. Though grown under a wide range of climates and soils, wheat is best adapted to temperate regions with rainfall between 30 and 90 cm. Winter and spring wheats are the two major types of the crop, with the severity of the winter determining whether a winter or spring type is cultivated. Winter wheat is always sown in the fall; spring wheat is generally sown in the spring but can be sown in the fall where winters are mild.

Therefore, today wheat is grown all over the world, with different varieties sown according to the various climates. In 2002, the world's main wheat producing regions were China, India, United States, Russian Federation, France, Australia, Germany, Ukraine, Canada, Turkey, Pakistan, Argentina, Kazakhstan and United Kingdom. Most of the currently cultivated wheat varieties belong to hexaploid wheat, which is known as common bread wheat and valued for bread making. The greatest portion of the wheat flour produced is used for bread making.

Wheat grown in dry climates is generally hard type, having protein content of 11-15 percent and strong gluten (elastic protein). The sticky gluten of bread wheat entraps the carbon dioxide (CO_2) formed during yeast fermentation and enables leavened dough to rise. The hard type of wheat produces flour best suited for bread making. The wheat of humid areas is softer, with protein content of about 8-10 percent and weak gluten. The softer type produces flour suitable for cakes, crackers, cookies, pastries and household flours.

Durum wheat (Triticum turgidum L.), which is the main tetraploid type, is also important, although its large, very hard grains yield low gluten flour that is the main source of semolina suitable for pasta, couscous, burghul and other Mediterranean local end-products. Apparently, no economically important diploid wheats are being cultivated as a crop anywhere in the world. Although most wheat is grown for human food and about 10 percent is retained for seed and industry (for production of starch, paste, malt, dextrose, gluten).

Wheat grain contains all essential nutrients; kernel contains about 12 percent water, including carbohydrates (60-80% mainly as starch), proteins (8-15%) containing adequate amounts of all essential amino acids (except lysine, tryptophan and methionine), fats (1.5-2%), minerals (1.5-2%), vitamins (such as B complex, vitamin E) and 2.2% crude fibers.

WHEAT GENOME

The haploid DNA content of hexaploid wheat (Triticum aestivum L. em Thell, 2n=42, AABBDD) is approximately 1.7 x 1010 bp, about 100 times larger than the Arabidopsis genome, 40 times that of rice and about 6 times that of maize. The large genome of bread wheat has resulted from polyploidy and extensive duplications, such that over 80% of the genome consists of repetitive DNA sequences. The average wheat chromosome is around 810 MB, 25 times that of the average rice chromosome.

This huge size arises from the origins of the wheat genome, of its triple structure (ABD genomes) and the inclusion of extensive duplications such that more than 85% of the whole genome is composed of repetitive, highly methylated sequences. Current estimates of gene number in higher plants vary between

25,000 and 43,000. According to current studies, effects of approximately 30,000 genes governed the wheat plant phenotype.

Drought Tolerance

World food production is limited primarily by environmental stresses. It is very difficult to find 'stress free' areas where crops may approach their potential yields. Abiotic environmental factors are considered to be the main source (71%) of yields reductions. Drought is one of the most common environmental stresses that affects growth and development of plants through alterations in metabolism and gene expression.

It is a permanent constraint to agricultural production in many developing countries and an occasional cause of losses of agricultural production in developed ones. Wheat production suffers from variability in yield from year to year and from location to location. One of the main environmental abiotic stress that is responsible for yield instability and limitations in wheat is drought stress, which may occur early in the season or terminally at grain filling and grain development.

Improvement of productivity of wheat cultivars under drought conditions becomes one of the important breeding programs objectives in wheat. Breeding for drought tolerant wheat cultivars is a major objective in arid and semi-arid regions of the world due to inadequate precipitation, shortage of irrigation water and high water demand for crop evapotranspiration in such climates.

Mechanism of Drought Tolerance

Drought tolerance in wild plant species is often defined as survival, but in crop species it is defined in terms of productivity. Rosielle and Hamblin defined drought tolerance as the difference in yield between stress and non-stress environments, while productivity is the average yield in stress and non-stress. A different definition regards drought tolerance as minimization of reduction in yield caused by stress compared to yield under non-stress environments. Also, it is defined as the relative yield of a genotype compared to other genotypes subjected to the same drought stress. Drought tolerance comprises drought escape, dehydration avoidance and dehydration tolerance mechanisms.

Drought escape

Drought escape through early flowering and/or short growth duration is advantageous in environments with terminal drought stress and where physical or chemical barriers inhibited root growth. On other hand, later flowering can be beneficial in escaping early-season drought, if drought is followed by rains. Under non-stress conditions, late-flowering varieties tend to yield higher than early-flowering ones. This is because the early-flowering varieties are likely to leave the yield potential unutilized.

Dehydration avoidance

Dehydration avoidance can be defined as the plant's ability to retain a relatively higher level of 'hydration' under conditions of soil or atmospheric water stress. Levitt recognized two plant types with respect to dehydration avoidance: plant that avoid dehydration by reduced transpiration 'water savers' and plants that use means other than reduced transpiration 'water spenders'. Important features are root characteristics, leaf and stomata characteristics and osmotic adjustment to lower the osmotic potential.

Dehydration tolerance

Dehydration tolerance describes the ability of plants to continue metabolizing at low leaf water potential and to maintain growth despite dehydration of the tissue or to recover after release from stress conditions. According to Hsiao and Boyer, translocation is one of the more dehydration tolerant processes in plants. It would proceed at levels of water deficit sufficient to inhibit photosynthesis. Ample information has been accumulated in the cereals to show that grain growth is partially supported by translocated plant reserves stored mainly in the stem during the pre-anthesis growth stages.

When water stress occurs and the current photosynthetic source is inhibited, the role of stem reserves as a source for grain filling increases, both in relative and absolute terms. Stem reserves may therefore be considered as a powerful resource for grain filling in stress-affected plants during the grain filling stage.

Model for Drought Tolerance

Many anatomical, physiological and biochemical traits are mentioned in the literature as being drought adaptive. This model will include those, which are currently considered of most potential value to the environment described, bearing in mind that not all traits are appropriate for all drought environments. The development of molecular probes for marker assisted screening of these traits would be an important objective, assuming their use is more efficient at identifying superior genotypes than conventional screening approaches.

Large seed size: Helps emergence, early ground cover and initial biomass.

Long coleoptiles: Helps for emergence from deep sowing. This is practiced to help seedlings reach the receding moisture profile and to avoid high soil surface temperatures which inhibit germination.

Early ground cover: Thinner, wider leaves (i.e. with a relatively low specific leaf weight) and a more prostrate growth habit help to increase ground cover, thus conserving soil moisture and potentially increasing radiation use efficiency. This trait would be more important in the Mediterranean type of drought environment where rain may occur during the early part of the cycle. It would be less useful in regions where the crop grows exclusively on stored soil moisture where dust mulching is practiced or where residue retention is practiced to avoid evaporation from the soil surface.

High pre-anthesis biomass: Potential for vigorous growth prior to heading provides the opportunity to take advantage of relatively good growing temperatures and moisture availability earlier in the cycle. Up to 40% of available water may be lost by evaporation directly from the soil in Mediterranean types of environments, so high early ground cover and biomass production may permit a more efficient use of soil water. Although most drought studies show that high water use efficiency (WUE) is not associated with better performance, ideally early biomass should be achieved with maximal water use efficiency to improve water availability during grain filling.

Good capacity for stem reserves and remobilization: Stored fructans can contribute substantially to grain filling especially when canopy photosynthesis is inhibited by drought. Traits that may contribute include long and thick stem internodes, with extra storage tissue perhaps in the form of solid stems. In studies where crosses where made between lines contrasting in the solid stem trait, the solid-stem progeny contained more soluble carbohydrate per unit of stem length, though total stem carbohydrate was unaffected due to narrower and shorter stems.

High spike photosynthetic capacity: Spikes have higher WUE than leaves and have been shown to contribute up to 40% of total carbon fixation under moisture stress. Awns contribute substantially to spike photosynthesis and longer awns are a possible selection criterion. While gas exchange measurement of spikes is time consuming and difficult to standardize, chlorophyll fluorescence should be considered as a more rapid means of screening for spike photosynthetic capacity under stress.

High RLWC/CTD during grain filling to indicate ability to extract water: A root system that can extract whatever water is available in the soil profile is clearly drought adaptive, but difficult to measure. Traits affected by the water relations of the plant, such as relative leaf water content (RLWC) measured pre-dawn or canopy temperature depression (CTD) during the day and C13 discrimination or ash content of grain or other tissues, can give indications of water extraction patterns.

Osmotic adjustment: Adjustment will help maintain leaf metabolism and root growth at relatively low leaf water potentials by maintaining turgor pressure in the cells. Some research suggests that the trait can be assayed relatively easily by measuring coleoptile growth rate of seedlings in polyethylene glycol.

Accumulation of abscisic acid (ABA): The benefit of ABA accumulation under drought has been demonstrated. It appears to pre-adapt plants to stress by reducing stomatal conductance, rates of cell division, organ size and increasing development rate. However, high ABA can also result in sterility problems since high ABA levels may abort developing florets.

Leaf anatomy: waxiness, pubescence, rolling and thickness. These traits decrease radiation load to the leaf surface. Benefits include a lower evapotranspiration rate and reduced risk of irreversible photo-inhibition. However, they may also be associated with reduce radiation use efficiency, which would reduce yield under more favorable conditions.

High tiller survival: Comparison of old and new varieties have shown that under drought older varieties over-produce tillers many of which fail to set grain while modern drought tolerant lines produce fewer tillers most of which survive.

Heat tolerance: The contribution of heat tolerance to performance under moisture stress needs to be quantified, but it is relatively easy to screen for.

Stay green: The trait may indicate the presence of drought avoidance mechanisms, but probably does not contribute to yield per se if there is no water left in the soil profile by the end of the cycle to support leaf gas exchange. It may be detrimental if it indicates lack of ability to remobilize stem reserves. However, research in sorghum has indicated that stay-green is associated with higher leaf chlorophyll content at all stages of development and both were associated with improved yield and transpiration efficiency under drought.

Mobilization in Wheat

Reserve accumulation

Reserve accumulation in the stem and the size of the storage strongly depend on the growing conditions before anthesis. Total stem nonstructural carbohydrate (TNS) at anthesis was shown to vary from 50 to 350 g kg^{-1} dry mass in different experiments. Under optimal growing conditions with regard to temperature, water regime and mineral nutrition, carbon assimilation rates are high and a proportion of the assimilation during stem elongation is reduced by stress, storage in stems is reduced.

For example, remobilized water soluble carbohydrate (WSC) were 641 mg and 1047 mg in water stressed and irrigated wheat, respectively, because the former had less storage than the latter. Under dry land field conditions only half amount of water-soluble carbohydrates was available for remobilization during grain filling, as compared with irrigated conditions. Stem length, as affected by height genes, is important in affecting stem reserve storage.

The Rht1 and Rht2 dwarfing genes of wheat were found to reduce reserve storage by 35% and 39%, respectively as a consequence of a 21% reduction in stem length. However, under the favorable conditions the advantage of the tall (rht) genotype in reserve storage was not expressed in greater mobilization to ear. Under these favorable grain filling conditions only about 20% of grain yield was contributed by stem reserves in all genotypes. The contribution of stem reserves to grain yield was greater in a tall than a short barley cultivar, but absolute yield was the same in both, indicating that the taller cultivar was lacking in current assimilation compared with the shorter one.

Reserve utilization

Stem reserve mobilization is affected by sink size, by the environment and by cultivar. The demand by grain yield sink is a primary factor in determining stem reserve mobilization. Environmental conditions that decrease current assimilation during grain filling pose an increased demand for stem reserves for grain filling. When wheat plants were shaded during grain filling, up to 0.93g of grain was produced per gram of assimilates exported from the stem.

Stem reserve mobilization is affected by water deficit grain filling. Even the rate of development of water deficit may affect mobilization. It is therefore to be expected that estimates of relative contributions of stem reserves to total grain mass per ear or to grain yield would vary among the different reports, according to the experimental conditions and cultivar used. These contributions were estimated to be anywhere between 6% and 100%.

It may be concluded that the reduction in current assimilation during grain filling, under different stresses, will induce an increase in stem reserve mobilization and utilization by grain. Drought conditions during grain filling often also involve heat stress, which reduces the duration of grain filling. There is normally an increase in the rate of grain dry matter accumulation under high temperatures, but it is not sufficient to compensate for the decrease in duration.

Stem reserves mobilization

In most wheat growing regions and especially in the Mediterranean grain filling is subjected to several physical and biotic stresses. Grain filling often occurs when temperatures are increasing and moisture supply is decreasing. Also foliar disease of wheat tends to spread and intensify towards and after flowering. Leaf rust, stripe rust and Septoria leaf blotch can result in total leaf destruction at grain filling. The common end result of all these stresses is the reduction in kernel weight, kernel shriveling, reduced test weight and loss in yield.

The current source of carbon for grain filling is assimilation by light intercepting viable green surfaces. This source is normally diminishing due to natural senescence and the effect of various stresses, in addition to the demand posed by maintenance respiration of live plant biomass. While root storage is important in some legumes and other species, there is no evidence that roots or leaves are as important as stems for reserve storage in the small grains. In most studies of stem reserves in the small grains, stems also include the leaf sheaths, which in themselves contain reserves.

In small grains, stems store carbohydrates in the form of glucose, fructose, sucrose and starch, but the main reserve is fructan. Storage is commonly analyzed as total non-structural carbohydrates (TNS) or water soluble carbohydrates (WSC). Fructan accumulated in internodes while they were extending, although most of the fructan in an internode accumulated after it was fully extended. When WSC was mobilized from the stem, the mass of fructose first increased then decreased, indicating that fructan was hydrolyzed at a faster rate than it is produced.

Methodology and selection for stem

An important form of drought stress tolerance is the tolerance to post-anthesis stress. In wheat, as in other cereals, grain filling depends partly on actual photosynthesis during this period and partly on carbohydrates stored during pre-anthesis which are translocated from vegetative plant parts. Under conditions of terminal drought (and heat) stress, net photosynthesis decreases significantly in wheat and the proportion of translocation of stored soluble carbohydrates aś a source for grain filling becomes larger. Stem reserve mobilization is affected by water deficit grain filling. Genetic variation exists within cereal crop species in the ability to sustain kernel growth by remobilization of reserves.

Blum et al. developed a technique for revealing translocation-based grain filling in absence of photosynthesis under post-anthesis stress in wheat. With this method, small plots are sprayed to complete wetting by a solution of magnesium or sodium chlorate (4% active ingredient), at the initial part of the linear phase of kernel growth, when final cell number of the kernel has been determined in each genotype. The chemical bleaches all plant surfaces upon contact without killing the plant. The desiccant kills all photosynthetic tissue including leaves, leaf sheaths, glumes and awns.

The treatment does not simulate drought stress. However, it simulates the effect of stress by inhibiting current assimilation. When the plants are devoid of their chlorophyll, grain filling can proceed only with translocated plant reserves. At harvest, 1000 kernel weight was compared between treated and non-treated plants, calculating the rate of reduction in kernel weight caused by treatment. The technique is being applied experimentally in Australia using potassium iodide as a desiccant.

The correlation across diverse genetic materials between the rate of reduction in kernel weight by chemical desiccation and the rate of reduction by drought stress was found to be significant. It was r = 0.81 *** and r = 0.79 ** over two years in Blum et al. (1983b) and r = 0.48 *** and r = 0.81** over two years in Nicholas and Turner. Chemical desiccation can be incorporated into breeding program in two ways. Firstly, it can be used to assess responses of individual advanced lines or families, always compared with non-treated controls under non-stress conditions. Secondly, it can be used in mass selection at the F_2-F_4 generation.

Estimate Drought Tolerance

Due to the complexity of drought tolerance in comparison to other traits, e.g., quality or disease traits, which are more clearly defined, efforts to increase drought tolerance could be achieved through selection for yield, which integrates all the unknown factors that will be important for improving drought tolerance. However, consideration should be given to whether the plants are grown in conditions which are representative of the target environments, genetic variation is maximum and the genotype x environment interactions are not too large.

Accordingly, many yield-based parameters were suggested to evaluate drought tolerance. Many of them were constructed in forms of indices, e.g., stress susceptibility index (SSI) suggested by Fischer and Maurer. The stress susceptibility index is the ratio of relative reduction in yield of genotype due to drought compared to the mean relative reduction in yield of all tested genotypes. This SSI is found to be equivalent to the ratio of yield under stress (Yd) to yield under non-stress (Yw), (Yd/Yw).

MOLECULAR MAPPING

Traditional methods of plant breeding have made a significant contribution to crop improvement, but they have been slow in targeting complex traits like grain yield, grain quality and abiotic stress such as drought. In traditional plant breeding, the plant breeder during selection of desirable plant from the segregation population faced the following problems:

— a large segregating population needs to be screened for a desirable trait e.g., yield and its component, quality, drought tolerance, disease resistance, etc.;

— wait for advanced generations F6 to start selection for quantitative traits, for which selection in early generation is not effective;

— it becomes very difficult to screen a segregating population for a desired trait when the trait is influenced by environment;
— contrasting forms are often not distinguishable at seedling stage, making it necessary to grow population up to the adult stage;
— it is difficult to undertake pyramiding of resistance genes, since selection of additional genes in presence of an existing resistance gene would be difficult.

To meet the great increase in food production necessitated by population growth and the higher standards of living expected by most of the developing countries, biotechnology brings new and powerful tools to plant breeders. One method receiving growing attention is the mapping of chromosomal regions affecting qualitative or quantitative traits. Polygenic characters, which were very difficult to analyse using traditional plant breeding methods, can now be tagged using DNA molecular markers.

Molecular markers allow geneticists and plant breeders to locate and follow the numerous interacting genes that determine a complex trait. Genetic linkage maps can provide a more direct method for selecting desirable genes via their linkage to easily detectable molecular markers. Combining marker-assisted selection methods with conventional breeding schemes can increase the overall selection gain and, therefore, the efficiency of breeding program.

With the use of molecular techniques it is possible to hasten the transfer of desirable genes between varieties and to introgress novel genes from wild species into crop plants. The plant breeder would like to exercise indirect marker aided selection (MAS) at the seedling stage in early generations, if possible. Availability of tightly linked molecular markers for a trait will facilitate such an indirect selection and help plant breeding by saving time and expense.

Molecular Marker Technologies

The development of molecular marker technologies during the last ten years has revolutionized the genetic analysis of crop plants. A significant progress has been made towards the use of molecular approaches in plant breeding. From the time of Gregor Mendel until the mid-eighties, morphological characters had been the major types of markers readily available for genetic mapping. Molecular marker technology has changed dramatically during the past two decades.

The first molecular markers were isozyme markers, which based on the different mobility of differently charged protein with the same enzymatic function on the gel. Enzyme markers have limited genome coverage and numbers. The term molecular marker is taken here to refer to markers identifying variation at the level of DNA, though biochemical markers such as isozymes have made a valuable contribution to the development of genetic maps in the late seventies and eighties for example, of tomato and maize.

The molecular markers, so developed, may be used for improving the efficiency of traditional plant breeding by facilitating indirect selection through molecular markers linked to genes for the traits of interest, because, these markers are not influenced by the environment and can be scored at all stages of plant growth. This saves time, resources and energy that are needed not only for raising large segregating populations for several generations, but also for estimating the parameters used for selection.

In addition to these applications, DNA markers can also be used for germplasm characterization, genetic diagnostics, study of genetic diversity, study of genome organization, etc. Molecular markers have already been used not only for the preparation of molecular maps but also for tagging genes, controlling traits of interest, for use in marker assisted selection (MAS).

In plant, using markers, several genetic maps were initially prepared in tomato. Subsequently, maps were constructed in different crops such as rice; maize, barley or wheat. There are two main types of molecular markers, isozyme markers and DNA markers. Markert and Moller were first to describe the differing forms of bands that they were able to visualize with specific enzyme stains and they were the first to introduce the term isozyme. Following this discovery further investigation showed that many of these enzymes were tissue specific and generally unaffected by environmental or other factors.

It was found that this type of markers has many limitations and disadvantages. The other type is DNA based markers. Recombinant DNA technology has provided new more powerful tools for studying genetic variation with a greater resolution than all previous experimental methods, including protein electrophoresis. Recombinant DNA technology can be applied to a variety of in vitro techniques, which include DNA isolation and production of new combination of heritable material by the splicing of the nucleic acids in vitro.

The main advantage of DNA based markers is that they give information of about any kind of sequence in the genome, not only of isozymes or highly expressed nonisozymatic proteins such as storage proteins but also unexpressed sequences. On the basis of the principles and methods employed, molecular markers can be broadly classified:- (i) hybridization based markers, (ii) PCR-based molecular markers and (iii) sequencing and DNA chip based markers.

Hybridization based markers

Now it is possible to gain information about the whole genome and any of its components, surpassing the limitations of protein electrophoresis, which only gives information on translated sequences. DNA based markers have overcome main limitations of protein electrophoresis since the detection of variation is not limited to coding regions and all categories of mutational events can, in principle, be detected.

In 1980, it was suggested that large numbers of genetic markers might be found by studying differences in the DNA molecule itself, revealed as restriction fragment length polymorphisms (RFLP). This technique involves restriction of genomic DNA followed by electrophoretic size separation of the fragments in a gel matrix. The fragments are then transferred to a membrane by Southern blotting and hybridized with a radioisotope labeled probe. Many sources of DNA can serve as probes, such as a small piece of genomic DNA, cDNA sequence or specific PCR products.

Sometimes, probes can be adopted from other species as heterologous probes for comparative mapping, which allows comparison of genome organization and evolution between the related species. The polymorphisms detected by RFLP rely on the specific and characteristic nucleotide sequence, which is recognized and cut by restriction enzymes. Insertions or deletions between restriction sites or mutations occurring at restriction sites result in length polymorphisms of restriction fragments. The range of fragment length and number of fragments depends on different enzymes. RFLP markers are often co-dominant and, therefore, very informative.

However, generating RFLP data is labor intensive and time consuming and requires a relatively large amount of DNA. RFLP maps have been developed for a number of species like maize, tomato, lettuce, rice, pepper, Arabidopsis, rye, barley and peanut. In wheat, due to low frequency of RFLP, this approach has been relatively less useful. This is sometimes attributed to polyploid nature, high proportion of repetitive DNA and large genome size. Despite these difficulties, sufficient applications of RFLP were practiced in wheat.

These purposes included, genome mapping, variety identification and marker aided selection. In wheat, RFLP's have been used to map loci for seed storage protein, flour colour, vernalization and frost resistance, dwarfing and vernalization, resistance to pre-harvest sprouting, tissue culture response, resistance against cereal cyst nematode, milling yield and different important agronomic characters.

PCR-based molecular markers

The development of new methods to perform analysis with molecular markers has been the focus of many recent studies and most of these are based on PCR amplification of genomic DNA. Polymerase chain reactions (PCR) have been considered to be the most revolutionary modern technique of molecular biology in 1980s. PCR is a powerful extremely sensitive technique with applications in many fields such as molecular biology diagnostics and population genetics.

Recombinant DNA techniques have revolutionized genetics by permitting the isolaticn and characterization of genes, allowing the detailed study of their function and expression during development processes, or as a response to environmental factors. More of the cloning methods involved can be accelerated and sometimes even circumvented by using PCR, and novel applications of the technique now permit studies that were not possible before.

The idea of PCR is a simple process in which a specific segment of DNA is synthesized repeatedly, resulting in the production of large amounts of a single DNA sequence starting from a minute quantity of template. The process depends on primer sequences of DNA, which match flanking sequences at both ends of targeted sequence. Through repeated denaturing, annealing and synthesized steps, the intervening sequence is synthesized in a 2n amplification.

Random amplified polymorphic DNA (RAPD): Random Amplified Polymorphic DNA (RAPD) is a PCR-based technique for DNA fingerprinting. This assay, unlike the PCR, does not require knowledge of the target DNA sequence, and a single arbitrary primer will support DNA amplification from a genomic template if binding sites on opposite strands of the template exist within a distance that can be traversed by the thermo stable equation usually random oligonucleotides (or 10 bases) used as primer to amplify discrete fragments of genomic DNA.

The primers are generally of random sequence and contain at least 50% G and C without internal inverted repeats. The products are easily separated by standard electrophoretic technique and visualized under ultraviolet (UV) illumination of ethidium bromide stained agarose gels. Polymorphism results from changes in either the sequence of the primer-binding site (point mutation), which prevents stable association with the primer or from changes, which alter the size or prevent amplification of target DNA (insertions, deletions or inversions). RAPDs can rapidly be used to construct linkage maps.

These advantages have allowed mapping of a wide variety of plant genomes. RAPD markers were also instrumental for map-based cloning of disease-resistance genes. However RAPD markers, inherited usually in a dominant manner, are not transferable from one population to another and are poorly reproducible between different laboratories. Consequently, RAPD is largely being replaced by a more robust DNA fingerprinting technique termed amplified fragment length polymorphism (AFLP).

DNA amplification fingerprinting (DAF): A modification of the RADP assay, named DNA amplification fingerprinting (DAF), has been described by Caetano-Anoles et al. The difference from the other procedures is that the PCR products are separated on polyacrylamide urea gels and visualized by silver stain. In the DAF procedure primers as short as five nucleotides, produce complex band patterns ideally suited for genome fingerprinting applications. Denaturing polyacrylamide gradient gel-electrophoresis (DGGE) has been used to resolve DNA sequence differences among fragments of similar or identical size. Using the DGGE procedure, single base differences result in altered migration of DNA fragments and thus produce polymorphic DNA fragments. Because of this the DGGE procedure is considered to be highly suitable for self-pollinating species.

Sequence-tagged site (STS): STS is a short, unique sequence that identified a specific locus and can be amplified by PCR. A pair of PCR primers, which are designed by sequencing an RFLP probe representing a mapped low-copy number sequence, characterizes each STS. PCR could be used to detect polymorphism in wheat with primer sequences derived from the á-amylase and ã-gladine genes. RFLP probe X*bcd*1231, linked with Pm4a locus was converted into an STS marker. Also, Roy et al. found an STS marker, which showed a strong association with pre-harvest sprouting tolerance in wheat.

Directed search (amplification of low copy DNA): In general, cereals have a high level of repetitive DNA sequences (about 70% of the DNA sequence in wheat are repetitive). Removal of repetitive DNA sequences before PCR has been reported to produce polymorphic and reproducible DNA fragments. Hydroxylapatite column chromatography is used to enrich low copy DNA sequences. Once the procedure is standardized, it could be useful to screen marker linked to disease resistance and other traits.

Sequence characterized amplified regions (SCARs): Paran and Michelmore developed a dependable PCR - based technique called sequence characterized amplified regions (SCARs). In this procedure, the polymorphic DNA fragment is cloned and sequenced.

Amplified fragment length polymorphisms (AFLPs): AFLP is a PCR based DNA fingerprint technique. It involves restriction of genomic DNA followed by ligation of adaptors to restricted fragments and preselective and selective PCR amplification of a subset of these fragments.

The amplified fragments are resolved on a sequencing gel and visualized either by autoradiography or fluorescent sequencing equipment, depending on the method of labelling or silver staining. In contrast to RAPD, AFLP is highly reproducible and also transferable between different populations. One major limitation of AFLPs is the dominant nature and the difficulty in identifying allelic variants at a specific locus although co-dominant AFLP markers have been found, however, in frequencies of 4-15 % among all polymorphic AFLP markers.

AFLP is now the first option to saturate a particular region of the genome when map-based cloning is applied to cloning target genes. Above all, one major application of AFLP is for molecular genetic mapping. It has been used to construct maps for barley, sugar beet, soybean, petunia, rice, wheat and tomato.

Simple sequence repeats (SSRs) or microsatellite: DNA sequences with di-, tri-, tetra- or penta-nucleotide tandem repeats are described either as microsatellites, as simple sequence repeats or as short tandem repeats. The first report of microsatellites in plants was made by Condit and Hubbell. These markers appear to be hypervariable, in addition to which their co-dominance and reproducibility make them ideal for genome mapping, as well as for population genetic studies.

Inter-SSRs are a variant of the RAPD technique, although the higher annealing temperatures probably mean that they are more rigorous than RAPDs. They are present in the vast majority of eukaryotic genomes. The total number of different dinucleotide blocks has been estimated for several species. The number of sites ranged from 103 to 105 depending on the species and repeat motif. Polymorphism produced by a variable number of tandem repeats has been demonstrated in a large number of species.

This feature has made microsatellites. a very attractive molecular marker for species with a narrow genetic base such as wheat and barley. This methodology is based on the use of primers complementary to SSRs. Multilocus profiles have been generated using different kinds of oligonucleotide containing simple sequence repeats as single primer or in combination with arbitrary sequence oligonucleotides.

These studies have shown the reproducibility of the patterns generated the Mendelian inheritance of the polymorphic amplified bands and their usefulness in the investigation of the genetic relationships. Mapping of this kind of markers in barley and rapeseed have been conducted. In wheat, Devos et al. searched sequence database and converted two microsatellite sequences into PCR based markers.

These markers were genome specific and displayed high levels of variation. More recently, a detailed genetic map of 279 microsatellite loci, another map of 50 loci, 65 loci for the D genome and have been developed for bread wheat. The availability of extensive molecular maps of wheat microsatellites will help in tagging genes of economic importance for marker assisted selection. In wheat, microsatellite markers have been used to tag several genes or quantitative trait loci (QTLs).

Sequencing and DNA chip based markers

Single nucleotide polymorphisms (SNPs) are single base pair positions in genomic DNA at which different sequence alternatives (alleles) exist in normal individuals in natural populations. Significant efforts towards large-scale characterization of SNPs were first initiated in human genome research. SNPs have since been shown to be the most common type of genetic variation in organisms.

Of all the different types of sequence change including single nucleotide substitutions, insertions/deletions and copy number variation in nucleotide repeat motifs, SNP represents about 90 % of human DNA polymorphism. SNPs have been found to occur with a frequency as high as 1 in every 202 bp in the mouse genome and 1 in 1000 bp in the human genome.

There are various methods for SNP detection and scoring. More commonly used are gel-based sequencing and high-density variation-detection DNA chips. SNP has many advantages and great potential for many applications.

Since SNPs exist over the whole genome of organisms with a relatively high frequency, they could facilitate the development of highly dense genetic function maps that would be highly valuable for genome analysis. Moreover, as the sequence context of the SNPs is already known, it has the potential for automation and can facilitate the genetic study of associating sequence variations with heritable phenotypes on a large scale. Because of this, there has recently been considerable interest in SNP discovery and detection for genome analysis of plant.

Different Types of DNA Markers

Several different types of DNA markers are currently available for genetic analysis and new marker types are being developed continuously. Markers differ from each other in many respects: the initial workload and costs for building up the marker system, running costs and ease of use, level of polymorphisms, inheritance, number of loci analyzed per assay, reproducibility and distribution on the chromosomes. Detection of polymorphism at the DNA level is usually based either on restriction patterns or differential amplification of DNA.

In order to make a choice from a number of molecular markers that have now become available, it is necessary to make a comparison between different types of molecular markers (RFLPs, RAPDs, DAF, SSRs, AFLPs, etc.). These markers have often been compared for convenience, reproducibility, speed of assay, cost effectiveness and feasibility of using automation and high throughput approaches. The choice of the best marker system depends on whether it will be used in evolutionary or population studies, genetic mapping or fingerprinting.

The ploidy level and reproductive system of the organism studied are also important. However, using suitably designed experiments in several crops including soybean, barley and wheat, these markers have also been compared for their relative effectiveness and efficiency in detecting DNA polymorphism for a variety of purpose. The superiority of AFLP and SSR techniques was also demonstrated in several other earlier studies. As mentioned above, the SSR markers have high information content are locus specific and co-dominant, thus making them the markers of choice for a variety of purposes including practical plant breeding.

However, if SSR markers are developed for the first time, a very high cost of their development will restricted their use in many laboratories. But if SSR primers have already been developed in a crop, the use of SSR markers for this specific crop will be cost effective and can be used even in small laboratories. Polymorphic bands in RAPD and AFLP can also be converted into locus specific SCARs, which will then prove to be as effective as SSR markers.

Mapping Populations

Mapping is putting markers (and genes or QTL) in order, indicating the relative distances among them and assigning them to their linkage groups on the basis of their recombination values from all pair wise combinations. Knowledge about the genetic concepts of segregation and recombination is essential to the understanding of mapping. The construction of a linkage map is a process that follows the segregation of molecular markers in a segregating population and put them in linear order based on pair wise recombination frequencies.

Thus, a mapping population with high number of polymorphisms over the total genome is highly desirable. Towards this end, various ways have been used to create mapping populations, which are illustrated in Figure 1. Populations used for mapping are usually derived from F_1 hybrids between two lines (either homozygous or heterozygous), which show allelic differences for selected probes.

Genetic maps of plants are constructed based on several different kinds of populations, with each population structure having unique strengths and weaknesses. Four types of population are commonly used for map construction and mapping experiment, they are F_2 population, back cross population (BC), doubled haploid (DH) population, and recombinant inbred lines (RILs). Most genetic mapping populations in plants have been derived from crosses between largely homozygous parents.

F_2 population

Such populations can be quickly developed and harbor all possible combinations of parental alleles. However, each F_2 individual has a different genotype and no replication or experimental design can be employed to effectively control environmental influence.

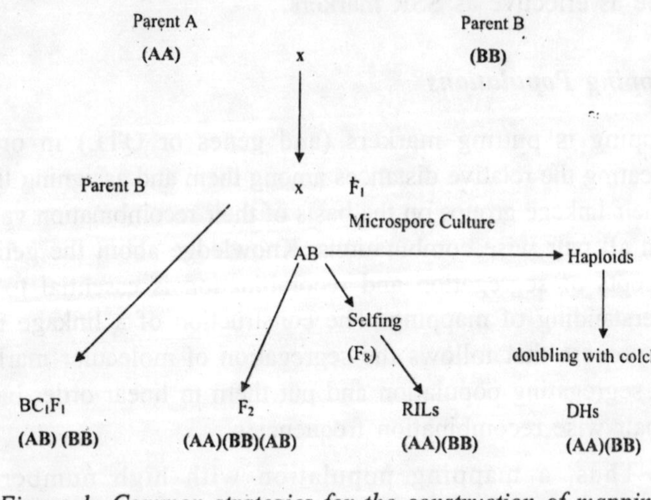

Figure 1: Common strategies for the construction of mapping populations with genotypes symbols (in bold).

To solve this problem, evaluation of F_3 progenies derived from individual segregants by selfing can be used but gains in precision are partly sacrificed due to genetic heterogeneity. A major disadvantage of F_2 population is that the data of marker genotypes cannot be repeatedly used.

Back cross (BC) population

This is widely used mapping population. It is derived by crossing F_1 individuals to one of the two parents and has similar advantages and drawbacks as F_2 populations. A major disadvantage of BC population is that the data of marker genotypes cannot be repeatedly used.

Doubled haploids (DH) population

A DH population can also be called a permanent population. Producing wheat haploids by crossing bread wheat with maize or pearl millet has become a significant procedure. Doubled haploids are commonly used in many plant species in recently, which are amenable to anther or microspore culture (from F_1 plants), followed by chromosome doubling. Because the plant has two identical homologues, the amount of recombinational information is exactly equivalent to a backcross. However, DH individuals are completely homozygous and can be self-pollinated to produce large numbers of progeny, which are all genetically identical.

This permits replicated testing of phenotypes and also facilitates distribution of identical DH populations to many different researchers. A major disadvantage of DH population is, it is not possible to estimate dominance effects and related types of epistasis and the rates of pollens or microspores successfully turned into DH plants may vary with genotypes, thus causing segregation distortion and false linkage between some marker loci.

Recombinant inbred lines (RILs) or single seed descent (SSD) population

Homozygous or 'permanent' populations can also be made by traditional means i.e., by selfing or sib-mating individuals for many generations starting from F_2 by the single seed descent (SSD) approach until almost all of the segregating loci become homozygous. Each of the loci having allelic difference in parents has two genotypes with equal frequencies. However genetic distances based on RILs population are enlarged compared to those obtained from F_2, BC or DHs populations, because many generations of selfing or sib mating will increase the chance of recombination.

Also, the RILs populations have several advantages, including reproduction, which favors the genetic analysis of quantitative traits because experiments can be replicated over years and locations; and the use of dominant marker types with the same efficiency as the co-dominant ones. A major shortcoming of RILs populations is that development of RILs population takes long time and it is not possible for all individuals to be homozygous at all segregating loci through limited generations of selfing or sib mating, which decreases the efficiency for QTL mapping to some extent.

Mapping Quantitative Trait Loci (QTLs)

Quantitative traits

Characters exhibiting continuous variation are termed quantitative traits. Quantitative traits show continuous phenotypic variation in a population resulting from the combined allelic effects of many genes and environmental conditions. In crop plants most traits of agricultural and economical significance exhibit quantitative inheritance, such as yield, plant maturity, disease resistance and stress tolerance. The genetic loci, which control quantitative traits, are referred to as QTL. QTL analysis has been a major area of genetically study for many decades.

For most of the period up to 1980, the study of quantitative traits has largely involved biometrical approaches based on means, variances, covariance of relatives and, consequently, very little was known about the biological nature of quantitative or natural variation in terms of number and location of the genes that underlie them. It is only during the past decade with the appearence of efficient molecular marker technologies and specific statistical methods that it became possible to follow the segregation of quantitative traits via linked markers and to detect effects, numbers and map positions of QTL.

Methods of QTL mapping

The identification of QTL for economically important traits has been achieved primarily by two approaches, either through linkage mapping to anonymous markers or through association studies involving candidate genes.

QTL analysis through a molecular marker approach: The principle of QTL mapping is to associate the phenotypically evaluated trait(s) with molecular markers using statistic tools. The map locations of QTL can then be estimated by the means of highly associated markers. Typically, the detection and location of the loci underlying quantitative trait variation involves three essential steps. First, a segregation population is created and characterized with molecular markers. This usually leads to the construction of a genome wide genetic map of the population.

Second, the individuals of the same population are phenotypically evaluated for the traits under investigation. Finally, genotypic molecular markers are analyzed for association with the phenotypic trait data using appropriate statistical methods. This type of QTL analysis can lead to the elucidation of QTL parameters in terms of number, position, effects and interactions between them. Association of morphological markers with quantitative traits in plants was observed early on and the first steps towards mapping of QTLs or polygenes were taken based on the scarce markers available.

Currently, complete genetical maps exist for many crop species and algorithms have been developed for QTL mapping in a wide range of pedigrees. The simplest methods were based on single marker analysis, where the differences between the phenotypic means of the marker classes are compared using F-statistics, linear regression or nonparametric tests. The computer program Mapmaker has been used extensively for performing interval mapping in plant studies. Interval mapping, now called simple interval mapping (SIM), searches for a single target QTL throughout a mapped genome.

QTL analysis through a candidate gene approach: The candidate-gene approach is a powerful and robust method. Compared to the genome wide mapping strategy, the chances of finding markers linked to putative QTL are maximized, since the selection of candidate gene markers is based on known relationships between biochemistry, physiology and the agronomic character under study. This approach has been applied successfully in various QTL analyses, such as mapping QTL for defense response to diseases in wheat, for resistance to corn earworm in maize and early growth traits in maize.

QTL mapping experiments for abiotic stress: In the traditional models of quantitative genetics simplifying assumptions were made about equality and strict additivity of gene effects. From the results of the QTL mapping experiments it has become clear that such assumptions are incorrect. In many mapping experiments, a relatively small number of QTLs accounts for very large portions of phenotypic variance, with increasing numbers of genes accounting for progressively smaller portions of variance, until the significance threshold is reached.

The proportion of phenotypic variation explained by each QTL and all QTLs together depends on heritability of the trait as well as on the portion of revealed QTLs. QTLs are usually spread over all chromosomes, but clusters of QTLs in certain chromosomal regions have been observed as well. Differences occur in QTL incidence when quantitative traits are scored in many environments or during many years. However, comparative studies between related species have revealed conservation not only in marker order but also in locations of some QTLs.

Molecular Markers and Its Application

The invention of molecular marker technology are RFLP, RAPD, AFLP, and SSR.It has opened up a new era for genetic analysis of plant genomes. Genetic mapping using molecular marker technology is of great significance to plant breeding, plant genetics and evolutionary studies. The most common applications of genetic linkage maps are concentrated on the following areas. First, genetic linkage maps can be used for marker-assisted selection (MAS) in plant breeding.

They could help to identify DNA markers linked to single genes of major agronomic importance and the tightly linked DNA markers can be used as diagnostic tools for MAS. This is particularly suitable and powerful for screening for monogenic disease resistance. One of the successful examples is MAS for soybean cyst nematode resistance (SCN). The SSR marker Satt309, which is located 1-2 cM away from the gene rhg1 for resistance to SCN, has been developed and used for tagging and tracking the gene through breeding programs, leading to the development of resistant lines.

The use of SSR markers has largely decreased the time and effort involved as compared to phenotypic selection. Second, genetic linkage maps can be used for the genetic analysis of quantitative traits. With the construction of molecular linkage maps, characterization of quantitative traits has been greatly facilitated in identifying the genomic regions responsible for the traits and estimating the possible number of genetic factors controlling the traits of interest.

Third, genetic linkage mapping can be used to correlate the phenotypic traits with the genes controlling the trait, which includes map-based cloning of a gene of known heritable phenotype and postulating candidate genes for a trait with known biochemical basis. Finally, genetic linkage maps provide insights into chromosomal organization and could be useful in map-based evolutionary studies by comparative mapping.

Linkage maps

Construction of a genetic linkage map is based on observed recombination between marker loci in the experimental cross. Segregating families, e.g. F_2 population or BC progenies, DHs population or RILs lines are commonly used. In barley the use of doubled haploid progenies produced from the F_1 generation simplifies genetic analysis. Doubled haploid lines have undergone only one meiotic cycle and carry a completely homozygous chromosome set. This means that the genetic information per plant is constant irrespective of the marker system used. Genetic map distances are based on recombination fractions between loci.

The Haldane or Kosambi mapping functions are commonly used for converting the recombination fractions to map units or centiMorgans (cM). The Haldane mapping function takes into account the occurrence of multiple crossovers but the Kosambi mapping function accounts also for interference, which is the phenomenon of one crossing-over inhibiting the formation of another in its neighborhood. Computer programs performing full multipoint linkage analysis include Mapmaker and JoinMap. Linkage map of human genome based on segregation analysis of 814 (CA)n microsatellite loci was initially constructed.

However in plants, mapping with STMS markers did not reach this level of resolution so far, although the very first attempt to map sequence-tagged microsatellite sites (STMS) loci in any species, was made as early as 1992, in rice using (GGC)n microsatellites. Several barley maps based on SSRs and randomly amplified SSRs have been developed. Mapping of the whole genome using microsatellite loci are also currently in progress in many crops i.e. Brassica, soybean and maize. Microsatellite loci, other than STMS markers, have also been used for mapping in different plant species.

In bread wheat, two microsatellite maps, one with 279 loci and another with 50 loci have been prepared. Also in tetraploid wheat, 14 microsatellite loci were mapped on chromosomes 5A and 5B, which carry genes for protein content, vernalization response and resistance to Hessian fly. Utilizing International Triticeae Mapping Initiative (ITMI) population, an integrated map of wheat genome became available, to which 279 gwm microsatellite loci were added.

Gene and Genome Mapping Group, Institut für Pflanzengenetik und Kulturpflanzenforschung (IPK), Gatersleben, Germany has successfully assigned a set of another 70 microsatellite loci to specific chromosomes using nulli-tetrasomic lines. Additional microsatellite loci have been mapped by Leroy. Comparative mapping within the Poaceae family has also revealed high levels of conservation of gene order.

Marker Assisted Selection (MAS)

Qualitative genes are inherited in a Mendelian fashion and their allelic forms give qualitatively distinct phenotypes. The phenotypes in a segregating progeny can be scored in a similar fashion as molecular markers. A normal segregation analysis will reveal linkages to any of the markers. Mapping a gene to a certain location on the chromosomes demands a linkage map of the whole genome, but genes can also be tagged with molecular markers without any previous information of the map location of markers used.

Two approaches have been proposed for this purpose, i.e., use of near-isogenic lines, NILs and pooled DNA samples. NILs differ only by the presence or absence of the target gene and a small region of flanking DNA. Hundreds of arbitrarily primed PCR-based markers can easily be screened to identify differences between isogenic lines and these differences are likely to be linked to the target gene. The NILs have been used in barley to tag a powdery mildew resistance gene and a spot blotch resistance gene.

In bulked segregant analysis (BSA), DNA pools of individuals of a crossing progeny are made based on their phenotype and screened for differences in the molecular markers. BSA has successfully been used in barley for tagging several disease resistance genes with RAPD markers locating 1.6-12 cM from the target locus. Also, BSA has been proposed for tagging quantitative loci with a major effect: theoretically QTL alleles with phenotypic effects of 0.75-1.0 standard deviations should be detectable in DH populations of 100-200 lines.

The first example of a gene linked to a microsatellite (AT) was a soybean mosaic virus resistance gene (Rsv). Several other resistance genes including those for resistance to peanut mottle virus (Rpv), Phytophthora (Rps3) and Javanese root knot nematode, were found to be clustered in the same region of soybean genome where this (AT)n microsatellite was found to be associated with Rsv. Microsatellite markers, associated with soybean cyst nematode (SCN) resistance locus, sclerotinia stem rot resistance and brown stem rot resistance.

In wheat, microsatellite markers have been applied widely for tagging genes or QTLs determining dwarfing, vernalization response, disease resistance, flour colour and milling yield, grain protein content, preharvest sprouting tolerance, grain yield and its components and frost. In durum wheat, some microsatellites have been mapped in two regions of chromosome 5A each carrying a QTL, for high grain protein content and for heading.

Variety identification and Fingerprinting

The ability to discriminate between and identify varieties of agricultural crops is central to the operation of seed trade. Plant breeder's rights offer protection for varieties, but in turn require that new varieties are Distinct from others, Uniform and Stable in their characteristics (the so called D, U and S criteria). Varietal identification and purity are also important for consumers, and especially for industry, which uses the harvested yield for large-scale processing.

DNA markers have been introduced as a promising method of fingerprinting different crop varieties. For example, DNA fingerprints of 65 registered sixrowed barley varieties in Canada have been generated using RAPD markers. All varieties could be identified from each other based on 18 polymorphic bands. AFLP and RAPD markers have also been successfully used for barley malt fingerprinting. Using a number of SSRs, oligonuclotide fingerprinting has been successfully utilized in a variety of crops. In a study involving 105 hexaploid wheat varieties from Argentina, closely related wheat varieties having high similarity co-efficients, could be discriminated using selected microsatellites, located on different chromosomes.

Genetic diversity

DNA polymorphisms can also be used to explore issues of genetic diversity. Knowledge of genetic diversity and the genetic relationship between genotypes is an important consideration for efficient rationalization and utilization of germplasm resources. Information on genetic diversity is also needed for the optimal design of plant breeding programmes, influencing the choice of genotypes to cross for the development of new populations.

Molecular approaches have been used to group barley cultivars into morphologically distinct groups and further into subgroups that have a similar genetic background. RFLPs, RAPDs, AFLPs and SSRs have been used for assessing variation in local and global collections of barley germplasm. Microsatellites have been considered to be the markers of choice for assessment of genetic diversity among cultivars and their wild relatives.

Using set of 12 microsatellitę markers, Prasad et al. studied the genetic diversity of 55 elite genotypes. In bread wheat, 42 microsatellite markers clearly distinguished varieties in three different agro-ecological areas namely Hungary, Austria and Germany. The usefulness of these markers for estimation of genetic relatedness has been demonstrated in many crops including maize, barley, wheat, rice, sorghum, sunflower, Aegilops tauschii and potato.

Genetic fidelity and germplasm

Microsatellite markers have been utilized to find out wether or not the germplasm accessions maintain their genetic fidelity during storage and conservation. One should realize that the failure to detect polymorphism might not indicate genetic fidelity, although reproducibility of polymorphism between the two taxa after storage may suggest some degree of genetic fidelity. The microsatellite have been used to characterize and conserve the germplasm in many plant species i.e. soybean , rapeseed and Phaseolus.

Chapter 4

Genome Sequence Database (GSDB)

A new era in biological research involving the comparison and functional analysis of complete genomes has begun. This revolution is the direct result of several technological advances. First, the advent of efficient, inexpensive, high throughput DNA sequencing strategies was necessary in order to produce large volumes of accurate sequences. The rate at which progress has been made in the Microbial Genome Initiative is evidence of the effect that high throughput DNA sequencing has had on biological research. In the last year the number of complete microbial genomes, that are available in the public nucleotide sequence databases, has more than doubled and another 100 are expected in the next few years.

Singly, each of these sequences and its associated biological annotation contribute to the advancement of the understanding of gene function and microbial biochemistry/physiology. Collectively the availability of these complete genomes provides researchers with raw data to perform comparative genomic studies. For example, a minimal gene complement for cellular life was determined based upon the comparative analysis of the Mycoplasma genitalium genome with that of Haemophilus influenzae. In addition, a comparative analysis (at the protein level) of eubacterial, archaeal, and unicellular eukaryotic genomes was conducted to investigate the origins of archaea and novel protein functions. These examples mark the beginning of the era of comparative and functional genomics.

The availability of complete genome sequences has not only had a significant impact on the study of microbes and the Human Genome Project, in terms of sequencing and mapping efforts, but also upon other areas of biology such as agriculture and bioremediation. The second technological advancement that was necessary for the realization of comparative and functional genomics was the development and improvement of sequence analysis algorithms.

Lastly, the development of effective computer and software systems to manipulate and manage large volumes of data was necessary for the realization of comparative and functional genomics. The ability to query the public nucleotide sequence databases in an efficient manner, when they contain more than one billion bases and associated biological annotation is the direct result of improvements in computer, database and network systems. The ability to produce, analyze, and store large volumes of nucleotide sequence and associated data are the technological cornerstones of comparative and functional genomics.

The Genome Sequence DataBase (GSDB) located at the National Center for Genome Resources in USA is poised to support many types of data that are essential to comparative and functional genomics studies. In August 1996 GSDB completed a database conversion that included upgrading of the Sybase system and improvement of the database schema. As a result of this conversion, GSDB can now store a complete genome (of any number of base pairs) as a single sequence, sequencing confidence data, sequence analysis results/scores, sequence alignments, discontiguous sequences (discontigs), data ownership and researcher defined features. While all of these changes are useful to the researcher community as a whole, these changes are crucial to the support of genomic level sequence comparisons and functional analyses.

In most public nucleotide sequence databases it is difficult to adequately evaluate the significance of a biological feature. For example, comparing a protein coding region that has been hypothesized in one sequence based upon significant homology to another sequence which contains a protein coding region is not possible given the data stored in most nucleotide databases. Therefore comparison of the two sequences can only be accomplished by repeating the homology search on the same data set using identical search parameters. Storage of search parameters and scores with hypothetic features would greaty diminish the need to recreate homology searches.

GSDB is the only public nucleotide sequence database designed to support the representation of such data. One of the primary driving forces behind both GSDB data related and programming projects are the needs of the research community. In order for a public database to be useful to researchers, the quality of data within it must be relatively high. In a relational database like GSDB improvements in data quality can occur from either the removal of erroneous data, or from the addition and creation of more meaningful data annotation and data relationships.

VECTOR CONTAMINATION

In this age of high volume sequence production, many researchers are relying more heavily on homology with annotated sequences in the public nucleotide sequence databases to determine the gene content of their sequence(s). Consequently, it is critical that the sequences in the public nucleotide sequence databases not contain vector contamination. The incorporation of vector sequences into the public nucleotide sequence databases has been a problem for a long time. Various researchers have examined this problem since 1992. Recently, the GSDB staff reassessed this data quality issue and analyzed the sequences in the database to determine the extent and progression of vector contamination.

The initial phase of the analysis involved the creation of a database that contained ~200 unique multiple cloning sites that are commonly used in cloning vectors. A homology search was then performed between each sequence in GSDB and the multiple cloning site sequence database. This initial screen was used to identify potentially contaminated sequences. The initial screening was limited to only the multiple cloning sites because these sequences are rather rare in natural sequences and they are frequently adjacent to the cloned sequence of interest.

Other regions of cloning vectors are frequently derived from naturally occurring sequences and they are infrequently adjacent to the cloned sequence of interest. By only searching against the multiple cloning sites the number of sequences that were incorrectly identified as containing vector contamination was minimized. This method of vector identification was tested on the data set that was available to Lamperti et al. in 1992. Following initial identification, a potentially contaminated sequence was compared to a database of complete vector sequences in order to identify the entire span of vector contamination.

In addition, these sequences were examined for the existence of a restriction enzyme site at the junction between the potential vector sequence and the cloned sequence of interest. The finding of a restriction site at the proposed vector-cloned sequence junction further supported the argument that all of the vector contamination in the sequence had been identified correctly. In the nearly 1 000 000 sequences that were examined, approximately 0.36% were found to contain vector sequences. Most of these sequences only contained vector contamination at either the 5' or 3' end, while a few contained vector contamination at both ends.

The vector contamination has been denoted with annotation in all of these sequences and they are in the process of being removed from the sequence and stored in a comment attached to the sequence. Over 75% of the sequences that were identified as containing vector contamination were also found to contain a restriction enzyme site at the junction. A much smaller number of sequences appeared to contain vector contamination not at the ends of the sequence, but at internal positions of the sequence. This potential vector contamination has been annotated as such, but will not be removed from the sequence until the data submitter has been consulted.

In the data submitted during the 5 year period that was examined, the overall level of vector contamination in the database appeared to remain relatively constant at a value of <1%. However, the study also revealed that >50% of the contamination that was incorporated into the database in the last 2 years was contained in EST and STS sequences. Since these sequences, especially the ESTs are routinely used in homology searches, it is imperative that the vector contamination be identified and removed from the database, or else research may be influenced by erroneous homology matches.

The immediate goal of identifying sequences within the database that contain vector contamination has been successfully completed and the procedure/program to remove this contamination is in development. It is anticipated that the subsequent removal of these vector sequences will be completed by May 1998. In addition, the automation procedure to identify vector contamination prior to its incorporation into the database is being designed.

One of the primary ways in which GSDB differs from the IC (International Collaboration-DDBJ, EMBL and GenBank) databases is in the types of data which can be stored in the database and in data representation. First, GSDB stores a contiguous string of nucleotides as a single sequence in the database regardless of size. Consequently a complete bacterial genome or fungal chromosome can be retrieved as a single file. In addition GSDB assigns ownership of data within the database to the contributing researcher.

Finally, GSDB supports the incorporation of many other types of data that are not supported in the IC databases. Included in this set of unique data types are: sequence alignments, sequence analysis data, sequencing confidence values, and discontiguous sequences. During the last year the GSDB staff has been utilizing these unique capabilities to represent sequence alignments and discontiguous sequences (Table 1) to augment sequence annotation and representation in the database.

SEQUENCE ALIGNMENTS

In the past 2 years several complete microbial genomes, viral genomes, fungal chromosomes, and naturally occurring plasmid sequences have been incorporated into GSDB. During 1997 the GSDB staff made significant progress in aligning small pieces of genomic sequences to the complete genomic sequence of the same organism. Base pair differences, as well as the base pair spans over which two sequences align are retained in the database. Approximately 2100 HIV 1 sequences were compared to the complete genomic sequence of HIV 1 strain HXB 2 the HIV research community reference standard sequence.

Alignments were created between this sequence and any other HIV 1 sequence that showed >80% homology over 95% of the length of the smaller sequence as determined by BLAST 2.0 and FASTA. A similar comparison was performed with the complete genome of Escherichia coli and other smaller E.coli sequences in the database. BLAST 1.4.9 and BLAST 2.0 were used to identify sequences that had >95% identity over the total length of the smaller sequence. In total, >2000 of the E.coli sequences within the database were incorporated into the complete genome record as sequence alignments. In this case, the base pair differences that occur between the complete genome and other aligned sequences are more likely the result of sequence variations or sequencing errors.

The GSDB capability of storing sequence alignment information is flexible in the kinds of sequence alignments that can be represented. In addition to representing intraspecies homology alignments, this capability can also be used to represent the relationship between a completed clone sequence and the subclone sequences were assembled into the complete clone. It can also be used to represent the relationship between a transcribed molecule (mRNA, rRNA, tRNA, etc.) and the corresponding genomic sequence. Lastly, it can be used to represent interspecies homology comparisons, especially those that are utilized to predict that a specific gene is in a sequence because of its similarity to a sequence with that gene from another organism.

Table 1: Sequence alignments

GSDB accession no.	Sequence description
GSDB:S:1649882	E.coli complete genome and sequence alignments
GSDB:S:43226	H.influenzae complete genome and sequence alignments
GSDB:S:135829	HIV complete genome and sequence alignments
GSDB:S:941927	M.janaschii complete genome and sequence alignments
GSDB:S:43082	M.genitalium complete genome and sequence alignments
GSDB:S:1492975	Rhizobium sp. pNGR234a complete sequence and alignments
GSDB:S:1025912	S.cerevisiae chrom. III complete sequence and alignments
GSDB:S:1650518	S.cerevisiae chrom. V complete sequence and alignments

GSDB:S:848927	S.cerevisiae chrom. VI complete sequence and alignments
GSDB:S:1327451	S.cerevisiae chrom. VIII complete sequence and alignments
GSDB:S:1637738	S.cerevisiae chrom. IX complete sequence and alignments
GSDB:S:1347770	Synechocystis sp. complete genome and sequence alignments

DISCONTIGUOUS SEQUENCES

A discontiguous sequence is a set of sequence fragments that are part of a larger contiguous sequence. (This set of sequences is provided a single GSDB sequence accession number, so that the set is easy to reference and retrieve.) Consequently there is a known physical relationship between these fragments, but they do not abut to one another, nor do they overlap. The known physical relationship between these fragments may be as simple as the knowledge that the fragments are all from the same cosmid (or any sequence of known size) and therefore cannot be more than X Kb apart.

Alternatively the relationship may be as specific as knowing both the order in which the fragments occur within the larger contiguous sequence and the physical distance between them. By constructing discontiguous sequences for sets of sequences in the database where physical relationships between the sequences are known, the GSDB staff has been organizing the data into more meaningful sets. One of the goals of the staff is to continuously review these sets and to add new fragments to the sets when they become available in the database. This effort will continue until all the gaps between the fragments are filled in and a single contiguous sequence has been 'built'.

In the past year the GSDB staff constructed two sets of discontiguous sequences, corresponding to the genomes of two agriculturally important crops, corn and rice. A single discontiguous sequence was constructed for each chromosome in each of these genomes.In addition, the discontiguous sequences that correspond to the human chromosomes 1-22 and chromosome X (15) were updated by the addition of a significant number of sequence fragments. The physical position of each sequence fragment is stored in the database in terms of kilobases from the left end of the discontiguous sequence and an uncertainty value.

In the case of the discontiguous sequences corresponding to the human chromosomes, most of the newly added fragments and all of the fragments that were used initially to construct these discontiguous sequences are STSs. The staff has also been working to place other, larger genomic sequences into the discontiguous sequences. This is being accomplished by performing BLAST homology searches between human genomic sequences and the sequence fragments that comprise a discontiguous sequence.

When a genomic sequence is found to overlap/contain a specific sequence fragment, the genomic sequence is incorporated into the appropriate position of the discontiguous sequence. In the upcoming year, efforts to improve data quality will continue both through the identification and correction of inaccurate data and data relationships, and through the incorporation of appropriate sequence relationships and other meaningful annotation.

Chapter 5

Gene Technology and Gene Ecology

Gene ecology is an emerging discipline. It is prompted by recent findings that the genetic material-DNA or in some instances, RNA-can transfer horizontally from one organism to another through the external environment, instead of vertically by reproduction. Gene ecology is the totality of how genes function, mutate, move, transfer and recombine subject to feedback regulation from the interconnected levels of the genome, the physiology of the organism and the external ecological conditions.

Gene technology, or genetic engineering, can profoundly disturb the ecology of genes. It is a technology for manipulating and transferring genes horizontally between species that do not normally interbreed. It is designed to break down species barriers and, increasingly, to overcome the species' defence mechanisms which degrade or inactivate foreign genes. For the purpose of manipulating, replicating and transferring genes, a range of artificial vectors have been constructed. Natural vectors for horizontal gene transfer comprise replicating, often mobile units of genetic material: viruses, plasmids and transposons. However, other pieces of DNA can also be taken up by cells and hence act as agent for horizontal gene transfer.

While viruses consist of genetic material wrapped in a protein coat, plasmids and transposons are, to varying extent, naked DNA. Plasmids generally replicate in the cytoplasm, while transposons are integrated into the chromosome, but can move from one site to another in the same or different chromosome, or from chromosome to plasmid and vice versa. Plamids and transposons typically carry virulence genes and genes for antibiotic resistance. Artificial vectors are made by joining together parts of the most infectious natural vectors in order to enhance horizontal gene transfer. The artificial vectors are 'crippled', which means that most, if not all of the genes causing disease and infectivity are removed. But it does not mean those functions cannot be supplied by, or regained from, other viruses and parasitic genetic agents that are always present in the environment and in the cells of all organisms.

The gene to be transferred is as a rule integrated into the genetic material of the vector; viruses, however, can also transfer genes that are not integrated, but merely packaged within the protein coat. To make a transgenic organism, the vector carrying the foreign gene to be transferred-transgene-is allowed to infect the cells of the organism. Once inside the cell, the vector can either multiply many copies of itself or become integrated into the genome of the cell. Most artificial vectors possess one or more antibiotic resistance marker genes, so transformed cells can be selected with the appropriate antibiotics.

There are three main routes for horizontal gene transfer: infection with viruses, through pieces of genetic material taken up into cell from the environment; or by unusual mating taking place between unrelated species. As the entire orientation of genetic engineering is to facilitate horizontal gene transfer, it is to be expected that antibiotic resistance genes as well as virulence genes will inadvertently spread and recombine to generate new, drug and antibiotic resistant pathogens. Indeed, the evolution of virulence and the spread of drug and antibiotic resistances are now linked to the extensive horizontal gene transfer and recombination events among bacteria and viruses, many of which may have occurred in recent years.

For example, horizontal gene transfer and subsequent genetic recombination have given rise to the new bacterial strains responsible for the cholera outbreak in India in 1992, and for the Streptococcus epidemic in Tayside in 1993. The *E. coli* 0157:07 strain involved in the recent outbreaks in Scotland is believed to have originated from horizontal gene transfer from the pathogen, *Shigella*.

Many unrelated bacterial pathogens, causing diseases from bubonic plague to tree blight, are now found to share an entire set of genes for invading cells, which have spread widely by horizontal gene transfer. Similarly, genes for antibiotic resistances have been transferred horizontally and have recombined with one another to generate multiple antibiotic resistances throughout the bacterial populations. Antibiotic resistance genes spread readily between human beings, and from bacteria inhabiting the gut of farm animals to those in human beings.

Antibiotic resistant strains of pathogens have been endemic in many hospitals for years. In the USA, up to 60% of hospital-acquired infections are antibiotic resistant. A special form of multiple which confer resistance to a wide range of chemically unrelated drugs made its appearance among pathogens in the early 1990s, although it has been produced in E. coli in the laboratory since the 1980s. Even more disturbing is the finding that antibiotics can increase the frequency of horizontal gene transfer 10-10000-fold. Thus, antibiotics create the very conditions facilitating the spread of antibiotic resistance.

The evolution and spread of antibiotic resistance predate genetic engineering, and are largely due to the profligate use of antibiotics in intensive farming, and in medicine. However, the abuse of antibiotics per se does not account for the emergence of new viruses and new virulent strains of bacteria. Fifty new viruses have been identified just between 1988 and 1996. The urgent question which needs to be addressed is the extent to which genetic engineering biotechnology, by facilitating horizontal gene transfer and recombination, is contributing to the resurgence of infectious, drug-resistant diseases, and will continue to do so if allowed to proceed unchecked. This question is all the more important in the light of yet other recent findings indicating that microorganisms genetically engineered for 'contained use' may not be effectively contained, while our regulators are contemplating further relaxations of the guidelines on contained use.

Thus, biologically 'crippled' strains of bacteria have actually been found to survive in the environment to exchange genes with other species. Naked viral DNA may be more infectious than the virus itself. DNA released from cells is not readily broken down in the environment, thereby retaining the ability to transform organisms; and routine chemical treatments for inactivating pathogenic microorganisms and viruses, before they are discharged into the environment, may be ineffective, leaving a large percentage of pathogens in an active, infectious state.

The combination of evidence is sufficiently compelling, especially in view of the precautionary principle, to warrant, at the very least, an independent public enquiry into the possible contributions of the industry to the etiology of infectious diseases.

ANTIBIOTICS AND EMERGING PATHOGENS

Mutation-random changes in the genetic material-and selection-survival of the fittest-were thought, by most biologists, to sufficiently explain all characteristics of organisms. If the mutation rates are very low, then 'selection pressure' would have to be very strong for the character to spread. Soon after antibiotics were used in medicine, it was discovered that livestock lived longer and grew faster when they were fed antibiotics to overcome infections. This led to 'prophylactic' treatments.

In fact, antibiotics were routinely fed to promote growth, to chickens, cattle, pigs and dairy cows, which also extended the shelf-life of meat, poultry, eggs and dairy product. In the 1970s. it was discovered that chickens given high doses of antibiotics resulted in resistant Salmonella strains in both the meat and the eggs. By 1983, mutant strains of antibiotic resistant Salmonella that attacked human beings had emerged. The emergence of new pathogens is strongly associated with mutant strains that cross species barriers. One of the clearest and most disturbing examples of mutant bacteria crossing from domestic animals to human beings, and of the transformation of a previously benign strain into a pathogen, is Escherichia coli, a common bacterium inhabiting the intestine of all human beings and many other mammalian species.

Most of the time, E. coli is harmless, which is why it has been the major tool for genetic engineers, who have used it and its plasmids routinely to clone genes since the 1970s. E. coli is the most intensively manipulated organism, and genes from species in every Kingdom have been transferred to and cloned in it. Perhaps it is not surprising that E. coli has emerged as a major pathogen. The most prominent among pathogenic strains, E. coli 0157: H7, appeared in 1982, which, far from being benign, caused dangerous haemorrhages of the colon, bowel and kidneys of human beings; and it broke out suddenly and simultaneously in several states in the USA.

Like most E. coli strains in the 1980s, it was moderately resistant to the antibiotics ampicillin and tetracycline. Since then, many outbreaks have occurred all over the world, and in increasing frequencies. A mass outbreak in Japan in 1996 affected 10000, with 12 deaths in children. A series of outbreaks in Scotland in 1997 claimed 20 lives and made hundreds ill. Genetic fingerprinting-a technique that fragments the DNA using specific enzymes and reading the fragments-backs up epidemiological evidence that E. coli 0157:H7 arose recently; and up to 1993, was endemic only to cattle in USA, Canada and Great Britain, and absent in cattle in the Far East. E. coli mutates fairly rapidly, hence the low degree of genetic variation observed among the 0157:H7 isolates before 1993 suggests they originated recently from a single clone. Most cases came from contaminated meat.

However, outbreaks have also been asso-ciated with cheese, salami, raw vegetables, unpasteurized apple juice and water. The E. coli 0157, like other water-borne pathogens such as Cryptosporidium, Giardia and Shigella, had developed resistance to chlorine. The mass outbreaks in Japan in 1996 were thought to be linked to white radishes, which may have come from contaminated soil. The specific toxicity of the strain is attributed to the Shiga-like toxin genes acquired by horizontal gene transfer from Shigella. One of these genes, VT1, is almost completely identical in base sequence to that of Shigella, indicating again, that the gene transfer was a very recent event.

Studies of dozens of emergent species of pathogens showed that genes for antibiotic resistance and virulence often reside in the same regions of the bacterial DNA, in plasmids or transposons, which may be passed horizontally from one species to another, picking up extra genes, recombining and generating novel combinations in the process.

Mechanisms for Antibiotic Resistance

A profusion of biochemical mechanisms and genes are involved in antibiotic resistances, fully matching, if not surpassing the variety of targets at which antibiotics are aimed. Most antibiotics have a single primary target, usually a single step in the synthesis of macromolecules essential to the bacterial cell, and preferably do not affect their eukaryotic host cells. Thus several classes of antibiotics are aimed at different steps in synthesizing the bacterial cell wall. These include the penicillins or β-lactams, vancomycin and ristocetin, D-cycloserine, fosfomycin and bacitracin. Other classes of antibiotics are aimed at bacterial enzymes involved in the synthesis of DNA, RNA and proteins.

These enzymes differ from their eukaryotic counterparts, although the effects of the antibiotics may not always be so selective, and are accompanied by varying degrees of toxicity for the eukaryotic host. Metronidazole, quinolones, nalidixic acid and novobiocin target DNA and DNA synthesis, while actinomycin and rifamycin inhibit transcription (RNA synthesis). Translation (protein synthesis) is targetted by a wide variety of antibiotics: streptomycin and other aminoglycosides such as kanamycin, neomycin, gentamicin, tobramycin, aikacin and netilmicin; tetracyclines, puromycin, chloramphenicol, erythromycin, lincomycin and clindamycin, and fusidic acid.

In addition, antimetabolites such as sulfonamides and trimethoprim inhibit the synthesis of essential metabolites. Resistances to every class of antibiotics have been identified (Table 1), the genes responsible may be found on the chromosome or in plasmids.

Table 1: A summary of mechanisms for antibiotic resistance

Mechanism	Antibiotic
1. Modification of targetto reduce or eliminate binding of antibiotic	β-lactams
	Erythromycin
	Lincomycin
	Streptomycin
2. Inactivation of antibiotic by enzyme	
a. Hydrolysis	β-lactams
	Erythromycin
b. Derivatization	Aminoglycosides

	Chloramphenical
	Fosfomycin
	Lincomycin
3. Sequestration of antibiotic by protein binding	β -lactams
	Fusidic acid
4. Metabolic bypass	Sulfonamides
	Trimethoprim
5. Binding of antibiotic by specific immunity protein	Bleomycin
6. Overproduction of antibiotic target	Sulfonamides
	Trimethoprim
7. Reduced uptake	Chloramphenicol
8. Active efflux from cell	Tetracycline
	Chloramphenicol
	Multiple resistance to broad ranges of antibiotics

One of the commonest mechanism of resistance is the degradation or modification of the antibiotic by specific all of which have arisen from one ancestral gene. On the other hand, aminoglycosides are inactivated by specific modifying enzymes encoded by 30 or more different genes, while chloramphenicol is inactivated by many acetyltransferases encoded by at least a dozen genes. Mutation of specific antibiotic target is also frequently involved in resistances.

For example, streptomycin is overcome in Mycobacterium by mutations in the ribosomal RNA (rRNA) molecules required for protein synthesis, so that they have reduced affinity for streptomycin. Antibiotics may be inactivated simply by sequestration, or binding permanently to specific proteins. A metabolic bypass may appear to a reaction blocked by the antibiotic, or the inhibited protein may simply be overproduced by the bacteria in order to overcome the inhibition. Finally, resistance may be conferred by reduced uptake of the antibiotics or by increased rates of pumping them out of the cell-active efflux mechanisms.

Active efflux mechanisms are especially important for the multiple resistance to broad ranges of structurally unrelated antibiotics and drugs that have appeared in pathogenic bacteria within the past 10 years. Quite often, several biochemical mechanisms are involved in giving protection against a single antibiotic, resulting in a very high level of resistance. Cross resistances are also common within a class of antibiotics such as the aminoglycosides, so at least one kanamycin resistance gene used as a genetic marker has been found to confer cross-resistance to two new generation aminoglycosides, amikacin and tobramycin. A recently documented case of antibiotic cross resistance is even more serious. Enterococcus faecium inhabiting the gut of turkeys on farms using avoparcin as growth promoter are found to have developed cross-resistances to vancomycin and teicoplanin, the last defence against multidrug resistant Staphlococcus aureus.

ORIGINS OF ANTIBIOTIC RESISTANCE GENES

Biochemically Specific Mechanisms: There are several sources of antibiotic resistance genes. The first is from organisms producing the natural antibiotic, such as penicillin, which also produce the inactivating enzyme. As new β-lactams are introduced, the genes encoding the β-lactamases undergo mutational changes, often involving only one base substitution and a single amino-acid difference in the enzyme. That would be sufficient to enable the enzyme to recognize the new drug and to break it down. In this way, the enzyme extends its spectrum of resistances to new drugs. Numerous mutational variants of β-lactamase have already been identified, showing the ease with which appropriate mutational changes can arise in this gene. The variants derive from different genes which are themselves traced to a single ancestral gene.

The versatility of the β-lactamase gene is such that when a series of inhibitors were made against the enzyme and used in combination with the β-lactam antibiotic-a kind of belt and braces approach-mutant enzymes arose, which not only break down the new antibiotic, but are, at the same time insensitive to the inhibitor. More often, however, it is not at all clear where the antibiotic resistance genes come from. New genes seem to appear from nowhere. For example, the 30 or more genes involved in resistances to different aminoglycosides are not related to one another. Some are thought to have arisen from mutations of pre-existing genes encoding enzymes catalyzing normal metabolic reactions in the cell. Others may involve activation of previously 'cryptic' or hidden genes in the bacterial genome. Or they may have been acquired by gene transfer from another bacterium.

Active efflux mechanisms: Active efflux mechanisms are due to 20 or more genes in at least 4 gene families, members of which can be found on plasmids or on the chromosomes. The genes code for transporter proteins situated across the cell membrane, that pump substances out of the cell, using ATP or protons as the source of energy. Recently, it has been shown that the genes for several efflux pumps are modulated in response to environmental conditions, and that many of the systems are also inducible by the substances to which they give resistance. The chromosomally encoded efflux genes in E. coli belong to one or another of four inducible operons, marRAB, soxRS, emrAB and acrAB. There is a great deal of homology among the genes comprising different operons.

For example, the proteins encoded by the regulatory genes, marR and emrR are two of the 15 proteins belonging to the

MarR family, members of which are found, not only in Gram negative bacteria inhabiting the gut, but also in Gram positive bacteria, and the mycobacteria, including plant and animal pathogens, as well as free-living species. One member of the MarR family, HpcR, found in soil-dwelling E. coli C strains, is a repressor of an operon encoding the enzymes that break down homoprotocatechuate (Hpc), a phenolic compound derived from plants. The repressor protein, HpcR. binds to the operator and represses the synthesis of enzymes that breakdown Hpc when Hpc is absent. But when Hpc is present, the repressor protein binds to it instead, freeing the operator, so that transcription and synthesis of the enzymes can takes place.

These observations suggest that naturally occurring noxious substances in the environment are the original inducers of the multiple antibiotic resistance operons. Further observations in support of the idea are that other naturally occurring compounds, such as salicylic acid, dinitrophenol and naphtho-quionones, stimulate the expression of both soxS and marRAB. The natural naphthoquinolones, plumagin and menadine, are effective inducers at relatively low concentrations, and mutants lacking both the marRAB and soxRS operons are notably hypersensitive to those compounds, while sensitivity to salicylic acid is only marginally affected. The marRAB operon expression is enhanced at 30ºC compared with 37ºC, as consistent with its origin in free-living bacteria.

The multiple antibiotic resistance operons have, therefore, evolved from free-living bacteria. The enteric bacteria have simply reactivated or mutated the genes in the operons to allow them to cope with drugs and antibiotics. This class of mechanisms is reminiscent of the multidrug resistance which all mammalian and plant cells are able to develop in the presence of sublethal doses of noxious substances.

RANDOM VERSUS DIRECTED MUTATIONS

To measure the rate of 'spontaneous' mutations to antibiotic resistance in a bacterial strain, a small amount of a dilute suspension of bacteria is spread on a nutrient agarplate impregnated with a concentration of the antibiotic high enough to kill nonresistant bacteria, so that only those with a pre-existing mutation which happens to confer resistance will survive and form colonies. The spontaneous mutation rate for antibiotic resistance estimated in this way is typically 10^{-9} or lower, and is about the same for all other genes in the genome. This is the rate of random mutations, in other words, mutations that have no direct correlation to the environment, or the 'selective' regime. Based on these estimates, microbiologists did not believe that resistance to antibiotics would pose a problem when antibiotics first came into use.

According to neo-Darwinian theory, is that it is due to the intense 'selection pressures' exerted by the antibiotics, so that even extremely low rates of spontaneous random mutations that happen to confer resistance to the antibiotic will be rapidly selected for, while those which do not happen to have the mutation will die out. Antibiotic resistance has been assumed to be a classic case of the natural selection of random mutations. What actually happens is something else. First of all, the mutations are not random in the sense that they pre-exist in the population, before exposure to antibiotics. All bacteria, with the exception of those that live in hot springs, are killed by boiling or cooking, which is therefore the surest way to prevent infection.

They can also be killed by high enough concentrations of antiseptics or antibiotics to which they are susceptible. However, at lower concentrations of the noxious agent, the bacterium may fail to grow at first, but after some time, develop the required resistance to the drug which enables it to grow, multiply and form colonies. This ability to develop drug resistance is common to cells of all species-bacteria, fungi, plant and animal without exception. It is a physiological response shared by all the cells or organisms in the population, and is not due to selection of pre-existing random mutations. It is similar to the 'directed mutations' or 'adaptive mutations', whereby starving, non-growing bacteria eventually acquire mutations that enable them to feed on new substrates.

Secondly, competition is irrelevant among bacteria in the evolution of antibiotic resistance: they share their most valuable assets for survival-genes encoding resistance mechanisms against antibiotics-by passing them on horizontally between species. In fact, the evolution of antibiotic resistance turns out to be a paradigm case of the 'fluid genome' in the new genetics, in which genes and genomes can mutate and undergo other changes when the ecological conditions call for them, and genes can even travel through the external environment from one organism to another.

Natural Genetic Engineering and Adaptive Mutations

Bacterial genomes are organized in modular fashion at successive levels of a genetic hierarchy. At the lowest level of the hierarchy are genes composed of different domains that are reshuffled and recombined to give different genes. Different combinations of similar genes give rise to multigene units or operons, and different combinations of operons make up larger units or 'regulons' for complex functions such as chemotaxis, symbiosis and protein secretion. Regulons may coincide with independently replicating 'replicons', such as plasmids and viruses, each with its own signal for replication.

Thus, mosaicism, or cut-and-splice genetic recombination is involved in every level of genome organization, and plays a key role in generating genetic diversity rapidly. But these mechanisms are by no means random. Bacteria may have 'little tolerance for purely random variability'. Bacterial cells possess a wide range of repair and proof-reading functions to remove accidental change to DNA sequences and correct errors resulting from physiological insults. But the same cells also possess numerous biochemical mechanisms for changing and reorganizing DNA, suggesting that such changes are the result of the 'natural genetic engineering' ability of bacteria to respond to environmental challenges.

The possibility of 'natural' genetic engineering mechanisms responding to the environment gained considerable support in the discovery of 'adaptive mutation', in which starving bacteria acquire the appropriate genetic mutation enabling them to metabolize a new substrate, when the substrate is present in the 'selective' medium. A wide variety of mechanisms are involved in generating adaptive mutations, including point mutations, deletions or insertions of nucleotides, conversions of gene sequences from one to another, site-specific integration of genetic material, DNA rearrangement and transposition. There is much current debate as to whether those mutations are specifically 'directed' by the environmental challenge, or simply the consequence of a hyper-mutable state of stressed cells that generates many mutations indiscriminately at a high rate, increasing the chances of a lucky hit. In either case, however, it means that genes and genomes change as the result of built-in mechanisms that are subject to metabolic and environmental feedback regulation.

Just as signaltransduction systems of the cell can direct the transcriptional apparatus to specific locations in the genome to activate those genes, so it is possible that the mutational apparatuses can be similarly directed to mutate specific genes. The evolution of antibiotic resistance and virulence as clear examples of 'natural' genetic engineering mechanisms at work, and goes no further than that. This can easily lead human genetic engineers to justify what they do on grounds that bacteria themselves have been at it, probably for hundreds of millions of years. That kind of reasoning is fallacious and dangerous. As some studies indicate, 'normal circumstances' for the vast majority of microbes in the environment as well as in our bodies may be a non-proliferating but metabolically active, non-virulent state. This state is associated with layers of extracellular matrices or 'biofilms' in which typically mixed communities of microbes are embedded, where they enjoy a cooperative, multicellular existence rather like the cells and tissues in our body.

Furthermore, there is nothing 'natural' about the 'manmade' environmental conditions that have provoked microbes to engineer their genes and genomes. The discharge of toxic wastes, the destruction of the environment, and the overuse and abuse of antibiotics, have already thrown the 'natural' ecology of microbes seriously out of balance. And over the past 20 years, the extensive manipulation of genes and genomes by human genetic engineers is very likely to have exacerbated the situation. The ecology of microbes consists, not only of the usually understood interrelationships with other organisms, but also of genes within the genome, of the genome within the physiological state of the cell and the organism, which is in constant intercommunication with the external environment.

Disease, cannot be understood apart from the ecological context. More importantly, it must be understood in terms of the nested levels in the ecology of genes and microbes, which is inextricably entangled with the ecology of human beings.

HORIZONTAL GENE TRANSFER

The major role of horizontal gene transfer in spreading antibiotic resistance has only been recognized since the late 1980s. The first definitive evidence for horizontal gene transfer came from DNA sequence analysis of the genes for neomycin-kanamyin resistance from Staphylococcus aureus, Streptococci and Campylobacter sp. which were found to be essentially identical. Identity of sequence not only gives evidence for horizontal gene transfer, but also indicates that the transfer is a very recent event. Further evidence of horizontal gene transfer was subsequently obtained for a wide range of antibiotic resistances, by DNA sequencing or other methods of characterizing DNA.

The scope of horizontal gene transfer is essentially the entire biosphere. Genes are found to have been transferred, not only among microorganisms and viruses, but among eukaryote species and across traditional Kingdoms of organisms. Among eukaryotes, the main transfers have involved transposable elements such as the P element among Drosophilids, and transposable introns among lower eukaryotes and bacteria. Some especially promiscuous elements such as mariner have been found in arthropods and vertebrates, including primates and humans, where it is associated with a neurological wasting disease.

Gene transfer mediated by transposable elements do not require DNA base sequence homologies. Transfers have occurred from bacteria to higher plants and vice versa. The best known example is the direct demonstration of transfer between the soil bacterium, Agrobacterium and plants. In a process bearing a strong resemblance to conjugation between bacteria, the tumour (T) segment of the tumour-inducing (Ti) bacterial plasmid is transferred and incorporated into the plant genome. However, it must be noted that nearly all cases of directly demonstrated horizontal gene transfer, especially those involving phylogenetically distant species, made use of already modified hybrid shuttle vectors that can transfer between species and replicate in both. These shuttle vectors possess signals for replication in more than one species as well as the signal for DNA transfer, and are hence much more likely to be successful in horizontal transfer than unmodified plasmids found naturally.

The Ti plasmid is, indeed, the basis of a gene transfer vector system widely used for genetically engineering crop-plants. Cross-Kingdom horizontal gene transfer by conjugation has also been demonstrated between bacteria and yeast using shuttle vectors derived from broad host-range promiscuous plasmids which are already transferable between many bacterial species. Such vectors can even substitute for the Ti plasmid in transferring genes from Agrobacterium to plants. Recent evidence documented the direct transfer of transgenes and marker genes from transgenic plants to soil fungi and soil bacteria, indicating that secondary, unintended gene transfers can occur from genetically engineered crop-plants which are now released commercially into the environment. Debris and exudates from transgenic plants will be expected to transform bacteria and other microorganisms in the soil. E. coli cells have been successfully transformed by T-DNA, simply using total DNA from a transgenic plant; and the T-DNA vector was recovered as a circular plasmid in the transformed cells.

In addition, other vectors such as viruses, insects and nematodes can also mediate unintended, secondary horizontal gene transfer. Mammalian cells are known to take up foreign genes by a number of mechanisms including anastomosis, endosymbiosis, phagocytosis and also endocytosis, a process dependent on specific cell-membrane receptors binding to DNA or to macromolecules complexed to DNA, so that the DNA can be internalized. It has even been suspected that gene transfer from bacteria to animal cells can occur by conjugation, a mating process previously thought to be confined to bacterial cells.

The ability of certain viruses to capture and transfer genes among mammalian cells is well-known and is associated with many animal cancers. Numerous illegitimate recombinations requiring no sequence homologies are due to retroviral insertion into the genome transferring genes horizontally between individuals of the same or different species. In short, a given gene from any organism has the potential to spread to all other organisms in the biosphere. However, DNA uptake by itself is not sufficient for successful gene transfer. Important barriers operate after the uptake of foreign DNA which break down, or otherwise block the integration, use and maintenance of the foreign genes. Plasmid DNA requires an origin of replication that is recognized and used by the host cell before it can be replicated and maintained.

Similarly, promoter sequences which can be recognized by the host are necessary for expressing the viral genes that make the virus, as well as expressing any other foreign genes carried by the virus. However, degraded DNA coding for parts or domains of proteins, or short regulatory motifs that bind transcription factors or otherwise enhance or repress gene expression may also be integrated, with significant genetic effects including cancer. Recombination of domains is already known to be a major factor in generating new proteins. While the host cell DNA has its own origins of replication, methylation patterns and promoter sequences which are specific to the species, artificial gene transfer vectors and other manipulated DNA have recombined origins of replication and promoter sequences from different species which can be recognized by different host species.

In addition, they are designed to overcome the restriction systems which break down and 'silence' foreign DNA. Thus, genetic engineering may have created new avenues and possibilities for increasing the scope of horizontal gene transfer. The scope of horizontal gene transfer among bacterial species has to be appreciated in the light of the revised phylogeny of the living world based on comparing ribosomal RNA (rRNA) sequences, a procedure first proposed by Carl Woese. Ribosomal RNAs are the most conservative molecules in evolution, and have not been subject to horizontal transfer between species. For these reasons, phylogenies based on rRNA sequences are generally considered to be the most reliable.

Comparisons of rRNA and other conserved sequences give a universal phylogenetic tree, revealing the true phylogenetic status of bacterial species. The tree is tripartite, with Bacteria (Eubacteria), Archaea (Archebacteria) and Eucarya (Eukaryotes) forming three more or less equal Domains. The traditional Kingdoms-Plants, Animals and Fungi -are but twigs on a small peripheral branch of the Eucarya Domain, each twig having the same phylogenetic status as traditional Orders of bacteria. Within the Bacteria Domain, E. coli is as distant from the Gram-positive bacteria, Bacillus, or Heliobacterium, as Homo sapiens is from the single-celled ciliate, Paramecium. But even the revised phylogeny of the living world greatly underestimates the genetic diversity of Bacteria. The known bacterial species are less than 1% of all species that exist, as more than 99% cannot be cultivated in the laboratory by routine techniques.

This highlights the dangers of releasing genetically engineered microorganisms or DNA into the environment. All known pathogens belong to the Bacteria and some to the Eucarya Domain. None belongs to Archaea despite the fact that some archaean species share our environment and are part of our normal bowel flora.

Gene Transfer Mechanisms

There are three main mechanisms for gene transfer in bacteria,

— by direct uptake of DNA (transformation)

— through genes being carried by viruses that infect bacterial cells (transduction) and

— through a mating process requiring cell to cell contact (conjugation).

These processes have been studied most extensively in E. coli, although it is becoming clear that what is true for E. coli does not necessarily apply to other species. The scope of our present ignorance is enormous, but what is known already gives us serious cause for concern.

Transformation among microbes in the environment has been extensively reviewed recently, showing that it is extremely widespread. The process involves DNA being released into the environment, its binding to the surface of bacteria, entry into the cell and recombination with the bacterial chromosome, or in the case of plasmids, reconstituting the plasmid. All these stages are facilitated by bacterial proteins. Both chromosomal and plasmid DNA are able to transform bacteria, the frequency of transformation varying with the physiological state of the cell, diffusible factors excreted from the bacteria and the presence of salts in the environment. In addition, the nature of the DNA, its source, size, and state may also be important.

Generally, transformation is most frequent if the DNA is from the same species, as its successful integration into the recipient genome depends on the degree of homology, i.e. the similarity in base sequence between the transforming DNA and the host DNA. It also depends on the DNA being resistant to the restriction enzymes in the host cell which break down foreign DNA. In one experiment, brief heat treatment dramatically increased the competence of cells for receiving foreign DNA by 4 orders of magnitude. This was attributed to the heat inactivation of restriction enzymes, thereby increasing the success of gene transfer. Cross species, cross-genera and even cross-order transfers have been observed with chromosomal DNA. Plasmid DNA, in particular, has effected cross-Kingdom transformations among Eubacteria, Proteobacteria and Cyanobacteria.

DNA is not only released into the environment when the cells die, but is actively excreted by living cells during growth. Some species export DNA wrapped in membranebound vesicles. The DNA in a culture slime can be more than 40% of the dry weight. Thus, the environment is extremely rich in DNA. Marine water contains between 0.2 to 44 mg per liter, and fresh water contains between 0.5 to 7.8 mg per liter; while freshwater sediment has an upper concentration of 1 mg per gram. Although enzymes breaking down DNA are found in the environment, DNA is protected from degradation by adsorbing to detritis, humic acid, and in particular, clay and sand particles. Adsorbed DNA is equally efficient in transforming cells.

Thus, the half-lives of DNA in soil is h for loamy sand soil, 15.1h for silty clay soil and 28.2 h for clay soil. While half-lives in waste water are typically fractions of an hour, those in freshwater and marine water are 3 to 5 h, with high values of 45 to 83 h on the ocean surface, and extremely high values of 140 and 235 h for the marine sediment. Adsorption of DNA to solid particles is a very rapid process, which means that DNA released into the environment can survive indefinitely and maintain its potential to transform other organisms. Transformation may be a major route of horizontal gene transfer: frequencies obtained under different environmental conditions, using artificial vectors and markers are found to be generally quite high, many ranging between 10^{-2} to 10^{-5} transformant per viable cell.

In fact, transformation frequencies are often higher under natural conditions than in the laboratory. In a recent study in soil microcosms, activity due to earthworms was found to significantly enhance gene transfer between spatially separated bacterial species innoculated into the soil. A special form of bidirectional transformation by cell contact and fusion is now known to be widespread. The release, or escape of naked recombinant DNA into the environment is a growing ecological problem, not only in the form of transgenic organisms, but also in vaccines, gene therapy vectors, and antisense oligonucleotide-short sequences of RNA complementary to the base sequence in the transcript of the genes-and ribozymes-RNA that act as enzymes to splice out intron sequences from transcribed genes-used in various forms of gene therapy.

Even short sequences can have significant genetic effects when integrated into the genome of organisms for which they are not intended. The hazard of naked DNA is highlighted by the discovery that some viral DNA may infect non-permissive hosts for which the virus itself is noninfective. Intravenous injection of naked, circular polyoma viral DNA into rabbits and mice gave a powerful genetic expression with the development of antibodiės and the production of new virus particles, whereas the native virus had no effect. Similarly, intravenous injections of recombinant plasmids in liposomes gave gene expression in several organs including the ovaries, indicating the possibility of an inherited genetic change. These findings are the basis for developing many recombinant DNA vaccines, for which ecological risk assessment has hardly even been considered. DNA vaccines are inherently dangerous due to their ability to infect non-permissive hosts, thus crossing wide species barriers, with many opportunities for recombination to generate new viruses.

Recombinant DNA vaccines are worse, as even small modifications of a virus can result in significant, unpredictable changes in its host range and virulence. In addition, they are more prone to recombine with other viruses to generate new viruses. Similar hazards are associated with the gene therapy vectors which are currently being developed.

Transduction is substantial in aquatic environments: Transduction takes place by means of bacteriophages which are usually host specific, and are hence not expected to transfer genes between unrelated species. However, promiscuous phages with broad host ranges are capable of transducing genes between unrelated species. A bacteriophage consists of viral DNA wrapped in a protein coat, which binds to specific receptor sites on the bacteria. Once bound to the receptor sites, the bacteriophage injects its DNA into the bacterium. The internalized phage DNA either directs the synthesis of many bacteriophages which eventually break open the cell, or the phage DNA may insert itself into the bacterial chromosome to become replicated with the chromosome.

The bacteriophage can transfer genes by mispackaging a piece of bacterial DNA in the viral protein coat, or by packaging a viral DNA that has a piece of bacterial DNA incorporated into it. When such a virus infects a second cell, the bacterial DNA is transferred and may then become integrated into the new bacterial chromosome. The transducing phage may serve as a vehicle for a transposon that can then establish itself in an unrelated host, even though the phage DNA cannot do so, and hence contribute to cross-species, cross-genera gene transfer. Also, bacteriophages themselves evolve by horizontal gene transfer and recombination, leading to broadening of their host ranges.

Phages are now known to be intimately involved in the horizontal transfer of entire blocks of virulence genes into the bacterial chromosome, turning a previously benign strain into a pathogen in a single step. Co-integrates of phage and plasmids are also involved in the trafficking of blocks of virulence genes.

Conjugation was previously thought to be a species-specific process requiring complex, complementary interactions between the genetic 'donor' and the genetic 'recipient'. Conjugation depends on certain conjugative plasmids, extrachromosomal pieces of DNA that are replicated independently of the chromosome. They carry genes required for the process of conjugation. These plasmids fall into different incompatibility groups. Conjugative plasmids can integrate themselves into the bacterial chromosome. In such cases, when conjugation takes place, the integrated plasmid will drag the bacterial chromosome with it across the conjugation tube from donor to recipient, leading to transfer of bacterial DNA as well as plasmid DNA. Donor DNA will then recombine with recipient DNA to generate new genetic recombinations.

The DNA is thought to be passed in a single stranded form, which is resistant to breakdown by enzymes attacking double-stranded DNA, thus ensuring a high success rate of gene transfer. The single-stranded DNA then directs the synthesis of the complementary strand to restore doublestranded DNA in both donor and recipient. Recent studies indicate that conjugation has an extraordinarily wide host range, involving diverse, complex mechanisms that are not yet well understood. 'Promiscuous' plasmids overcome species barriers, transferring genes promiscuously between phylogenetically distant species. At least several of the promiscuous conjugative plasmids have been exploited in constructing artificial shuttle vectors between E. coli and other distant species. One promiscuous conjugative plasmid is routinely used in Agrobacterium-mediated gene transfer to make transgenic plants.

Shuttle vectors are essentially unstoppable, as they typically carry the origins of replication for both species as well as the origin of conjugative transfer. The strength of conjugative transfer is in the single-stranded DNA transferred directly, which escapes destruction by restriction enzymes. While the shuttle vectors are constructed to require 'helper' transfer functions provided by other plasmids, those plasmids can be present in the donor or recipient, or in a third species altogether. These 'tripartite' matings can easily occur in all environments, which are full of known and especially, unknown bacterial species. In addition to conjugative plasmids, conjugative transposons mediate their own transfer to recipient cells during conjugation and insert into the recipient chromosome. They can also jump from one site to another on the same chromosome, or from chromosome into plasmids and vice versa. These have also been exploited in constructing gene transfer vectors.

A special class of such conjugative transposons, 'integrons' are thought to be especially involved in the evolution of the multiresistance R plasmids within the past five decades. An integron carries its own gene encoding an enzyme, integrase, that catalyzes the site-specific integration of antibiotic resistant gene 'cassettes' within the integron, such that each inserted 'cassette' is provided with a ready-made promoter for expression. An integron can carry several cassettes, each encoding a different antibiotic resistance. The integron can facilitate recombination between cassettes to form exotic gene-fusions that code for multifunctional proteins.

Integrons can also jump from the bacterial chromosome into plasmids and become transferred to another bacterium during conjugation. During conjugation, genetic material not only goes from donor to recipient, but also in reverse, from recipient to donor, in a recently discovered process, retrotransfer.

Environmental pollutants: Horizontal gene transfer is sensitive to many physical and chemical factors. In routine laboratory procedures for genetic transformation, heavy metals, singly or in combination, are used to greatly increase the competence of cells for transformation, as are enzymes such as muraminidases or proteases. Freezing and thawing, treatment with EDTA or other chelating agents, exposing cells to electric fields, or to DNA wrapped in liposomes, are all effective in increasing the frequency of transformation. Yet, very few experimental investigations have been carried out into the effects of environmental pollutants on horizontal gene transfer. This is a glaring omission in view of the hundreds, if not thousands, of xenobiotics that currently circulate in our environment.

And to which genetic engineering biotechnology is releasing, or discharging large volumes of agents-transgenic plants and animals, microorganisms, vectors, vaccines, and a range of naked DNA and oligonucleotides -all capable of taking part in horizontal gene transfer. Xenobiotics can affect horizontal gene transfer in two ways. Some xenobiotics mutate DNA, thus influencing its uptake into cells. Examples of minor changes in a DNA sequence which alter the host range of a transferable genetic element. Other xenobiotics can affect the cell membrane and:or intracellular functions thereby influencing the ability of cells to take up DNA, or to engage in conjugation and other forms of gene exchange.

Antibiotics can increase horizontal gene transfer 10 to 10000-fold. Antibiotics are heavily used in medicine and intensive agriculture, especially in fish farming, and constitute an important class of xenobiotics that already affect horizontal gene transfer significantly. It is known that ancestral antibiotic resistance genes have originated from antibioticproducing microorganisms themselves. The possibility has been suggested, therefore, that what we call antibiotics may actually be sex hormones for the bacteria, enhancing conjugation between cells.

The process is naturally regulated by the product of the 'antibiotic resistance' gene, which shuts off the signal.

Thus, the use of antibiotics not only leads to directed mutations of antibiotic resistance genes, it promotes the spread of those very genes among pathogens. The case for phasing out the use of antibiotics in intensive agriculture and fish farming is overwhelming. Antibiotics are also routinely used in genetic engineering biotechnology to select for transformed cells and recombinants. The importance of synergistic interactions between xenobiotics is increasingly recognized. Recent research has shown that chemicals that are apparently harmless when tested singly, can have enhanced effects in mixtures at very low concentrations. As xenobiotics never occur singly in the environment, it is necessary to test drugs and chemicals in combination for realistic risk assessment. But this is a logistically and financially impossible task.

The synergistic effects of xenobiotics on horizontal gene transfer are, for the most part, unknown and unpredictable. The precautionary principle demands that we should strive to limit the release:escape of genetically engineered organisms, naked DNA and RNA sequences into the environment as much as possible.

Horizontal Transfer of Antibiotic Resistance Genes

Antibiotic resistance genes, especially those carried on R plasmids and transposons, can in principle, cross species as well as genera and Kingdoms and even Domains. Evidence for transfer is obtained wherever and whenever it has been looked for, either by sequence analyses or by direct experimental investigations. Examples are given below.

Tetracycline resistant genes are now found to be shared between many genera of bacteria, one of them, the TetL gene, is found in Actinomyces, Clostridium, Enterococcus, Listeria, Peptostreptococcus, Streptococcus, Staphylococcus and Bacillus.

Another gene, the TetK, is found in Bacillus, Clostridium, Enterocococcus, Eubacterium, Listeria, Peptostreptococcus, Staphylococcus and Streptococcus. More recently, both TetL and TetK have been found for the first time in soil bacterial species belonging to the genera Mycobacterium and Streptomyces, which can cause soft-tissue and skin infections.

Antibiotic producing Streptomyces are believed to be the ancestral source of many aminoglycoside antibiotic resistance genes, but it is now gaining more of these genes back from the other species. These findings are of particular significance, as they suggest the potential for the spread of an antibiotic resistance gene into all environmental mycobacteria, including Mycobacterium leprae. Even chromosomally encoded penicillin-resistance genes are readily transferred. Transfers of pbp genes were found between Streptococcus pneumoniae and Neisseria gonorrhea which subsequently recombined to generate new hybrid genes.

Horizontal Gene Transfer Invoved in the Emergence of New and Old Virulent Straints

The very same genetic mechanisms for horizontal gene transfer have been involved in the emergence of virulence among old and new pathogens since the mid 1980s. DNA sequence analysis of virulence factors suggests that they may have been recently acquired by horizontal gene transfer. Pathogenic mycoplasmas possess proteins called adhesins and related accessory proteins which are required for adhering to cells and subsequent disease development. Mycoplasma genitalium, implicated in urethritis, pneumonia, arthritis and AIDS progression, was found to encode one adhesin and 2 adhesin-related proteins that shared substantial sequence similarities, as well as genome organization, with those of another species, M. penumoniae.

These were attributed to horizontal gene transfer events between the two species. Most disturbing of all is the recent discovery that at least 10 unrelated bacterial pathogens, causing diseases from bubonic plague to tree blight, share an entire set of genes for invading host cells, which have spread by horizontal gene transfer, as they reside as a block, often in the bacterial chromosome, forming 'pathogenicity islands'.

More than 20 genes are involved in a system that secretes damaging proteins directly into host cells. They were first discovered in Yersinia, a genus of several species that cause human disease including bubonic plague and intestinal infections. At least some of the proteins are typical of eukaryotic cells, and are targeted at eukaryotic biochemistry, leading to the speculation that the genes were originally acquired from eukaryotic cells.

Like antibiotic resistance genes, many virulence genes reside on transposons or plasmids. For example, the ST-enterotoxin genes of E. coli are part of a transposon; while important virulence factors of Gram-negative pathogens as well as Gram-positive pathogens are encoded in plasmids. Other factors, including the Shiga-like toxins of E. coli, the cholera toxin of Vibrio cholerae, diphtheria toxin of Corynebacterium diphtheriae, neurotoxins of Clostridium bolulinum and cytotoxin of Pseudomonas aeruginosa, are encoded by bacteriophage. All these toxins and virulence factors obviously have the capacity to spread horizontally and to create new pathogens by recombination. Virulence factors are often also located on the chromosome, where they form large blocks, which came to be called pathogenicity islands.

Pathogenicity islands: The term 'pathogenicity island' (Pai) refers to large chromosomal regions in pathogenic bacteria that encode virulence genes acquired by bacteria from unrelated organisms. A Pai according to the following criteria:

— Carriage of virulence genes.
— Presence in pathogenic strains, and absence or sporadic distribution in less-pathogenic strains of one species or a related species.

— Different G+C content in comparison to DNA of host bacteria, indicative of a foreign source.

— Occupation of large chromosomal regions (often more than 30 kb).

— Represents compact, distinct genetic units, often flanked by direct repeats.

— Association with tRNA genes and:or insertion sequence (IS), a special class of mobile genetic elements, at their boundaries.

— Presence of (often cryptic) mobility genes (IS elements, integrases, transposases, origins of plasmid replication).

— Instability.

A bacterial pathogen may have one or more pathogenicity islands. For example, uropathogenic E. coli strain 536 has two Pais: Pai I encodes only a-haemolysin, responsible for haemolysis, while Pai II carries the a-haemolysin gene, (hly), as well as the P-related fimbriae gene (prf) involved in attachment of the bacterium to the host cell. Pai I is flanked by the selC gene encoding the selenocysteine-specific tRNA and two direct 16 bp repeats, while Pai II is associated with leuX, which encodes a leucyl tRNA, and two direct 18 bp repeats. The direct repeats in both cases are attachment sites for bacteriophages. Cryptic integrase genes of these phages are located near the respective tRNA genes.

Thus, the Pais probably originate from integrated prophages which have picked up a range of virulence genes to turn them into pathogenicity islands. In addition, sequences homologous to the origins of replication of plasmids, parts of IS elements, and 'recombination hot spots' are also present. All these indicate that Pais are highly mosaic, and many recombination events have contributed to their evolution. Phages, plasmids and pathogenicity islands are involved in rapid evolutionary changes of bacteria by 'quantum leaps'. Pais often exhibit features of both plasmids and phages, reflecting the formation of co-integrates of both types of extra-chromosomal elements. The acquisition of new genes following horizontal gene transfer results in the generation of new variants of pathogens.

Drug-resistant strains of Mycobacterium tuberculosis and Streptococcus penumoniae, the new V. cholerae serotype 0139, E. coli 0157, Haemophilus influenzae biotype aegypticus, are examples of such new types of pathogens generated by horizontal gene transfer. Enteropathogenic bacterial strains, including E. coli 0157:H7, have the ability to remodel the actin cytoskeleton of infected cells, producing so-called 'attaching and effacing' (AE) lesions in the host cell. A 35.4 kb pathogenicity island in enteropathic E. coli, called the locus of enterocyte effacement, (LEE), has been found to contain all the known genes necessary for AE lesions. Several features of the LEE suggest that it has been acquired recently by the enteropathogenic E. coli and other AE bacteria. These include,

— a G+C content of 38.4%, compared with the 51% of the bacterial chromosome as a whole.

— the strict conservation of LEE base sequences within AE pathogens of 3 genera of bacteria, and

— insertion within selC tRNA gene, the same site of insertion for paiI in uropathogenic E. coli and the bacteriophage fR73. The LEE is one among a whole class of genetic systems that are extremely sophisticated in exploiting host cell functions.

Exploitation of host cell functions by bacterial pathogens: Recent studies reveal the common strategies used by a wide range of pathogenic bacteria as well as the unique tactics adopted by individual species in establishing infection. Many host cell functions are exploited by the pathogens, including signal transduction pathways and rearrangements of the cytoskeleton. Adhesion to host cells is the first stage in disease development. Bacterial pathogens produce a range of proteins, adhesins, which mediate adhesion to host cells. These adhesins interact with surface membrane components of mammalian cell, many of which serve as receptors for the bacterial adhesins. The host cell is often an active participant in adhesion by synthesizing the receptors for adhesion.

Among Gram-positive bacteria, Listeria monocytogenes enters mammalian intestinal cells by means of a surface protein, internalin. Internalin contains multiple copies of a 22-amino acid leucine-rich tandem repeat, a feature of several eukaryotic proteins that are generally involved in protein-protein interactions. Another surface protein, InlB, belonging to the internalin multigene family and displaying similar LRRs, mediates entry into cultured hepatocytes, Hela cells and Chinese Hamster Ovarian cells.

Barriers to horizontal gene transfer: The frequency of horizontal gene transfer may therefore be determined, not so much by the actual mechanisms that exist for mediating transfer as on the existence of barriers that act against the process. In bacteria, a major mechanism for limiting horizontal gene transfer is restriction-the destruction, or inactivation of foreign DNA. There are 11 restriction:modification systems in enteric bacteria which fall into three families, hsdR, hsdM and hsdS. The alleles of the hsdR family are responsible for destroying foreign DNA, while those of hsdM and hsdS families are responsible for methylating, thereby protecting DNA molecules that escape restriction and facilitating their recombination into the genome.

A second barrier is the mismatch DNA repair system encoded by the mutL, mutS and mutH genes of E. coli and S. typhimurium, which severely inhibits recombination of nonhomologous DNA. However, another genetic system, the error-prone SOS system, activated under stress, stimulates such recombination. In addition, there are origins of replication and origins of transfer specific to DNA from different species, so that even if DNA is transferred, they cannot be replicated and maintained, so that the chances of integration into the chromosome would also have been much reduced, unless, of course, the DNA has been manipulated to combine origins of replication from different species. Another important influence on the frequency of horizontal gene transfer is the ecology of the microbes concerned..

Stressful conditions appear to reduce the effectiveness of restriction and to encourage recombination. Starving bacteria are also more competent in taking up DNA from the environment, and hence to become transformed. And heat treatment greatly increases the frequency of transformation; while the effects of xenobiotics in the environment on horizontal gene transfer, apart from antibiotics, remain largely unknown.

One main contributing factor to the recent increase in the scope and frequency of horizontal gene transfer may be the deliberate acts of genetic engineers to break down species barriers. They do so by constructing a range of chimaeric vectors for cloning and transferring genes. These artificial vectors have the following important characteristics that enhance horizontal gene transfer.

— They are already derived from elements that mediate horizontal gene transfer most effectively.
— Their chimaeric nature means that they possess sequence homologies to DNA from widely different species and their viral pathogens, plasmids and transposons, thus facilitating successful horizontal transfer and recombination.
— They routinely contain antibiotic resistance marker genes enhancing their successful transfer in the presence of antibiotics, either intentionally applied or as xenobiotics in the environment.
— They often have origins of replication and transfer sequences, all of which facilitate horizontal gene transfer and recombination. Here the fact that they are 'crippled', so that genes for mobility and:or virulence are removed, is irrelevant, as helper functions can be supplied by other viruses, plasmids and mobile genetic elements present in

the donor, recipient or a third strain of bacteria. And virulence genes can be regained by recombination.

— It is well-known that chimaeric plasmids and viral vectors are subject to structural instabilities which make them more prone to recombine. Vector instability is a continuing problem for genetic engineers and the biotech industry as far as the stability of the transferred genes is concerned. It also increases the probability and scope for unintended, secondary horizontal gene transfer, which has already been directly demonstrated.

— The now routine incorporation of strong promoters and enhancers in vectors to boost expression of transgenes is one main cause of structural instability, which is in addition to the instability arising from the attendant metabolic stress to the organism, that again, may increase unintended horizontal gene transfer.

— Finally, vectors are designed to escape restriction thereby also enhancing the probability of successful horizontal gene transfer.

The construction of artificial vectors is fundamental to genetic engineering. All classes of genetic elements that mediate horizontal gene transfer have been used in constructing vectors: plasmids, phages, transposons, plus a range of pathogenic plant and animal viruses. As stated in a standard textbook on genetic manipulation, 'Many animal viruses have been subjugated as vectors. Virtually every virus that has been studied in any detail and that has a DNA genome or a DNA stage in its replication cycle has been manipulated in this way'. Although different classes of vectors are distinguishable on the basis of the main framework sequence, practically everyone of them is chimaeric. Important chimaeric vectors are the shuttle vectors which enable genes to be cloned in E. coli and transferred into unrelated species in every Kingdom.

Similarly, vectors used in manipulating plants and animals typically contain sequences from a range of plant and animal viral pathogens, as well as antibiotic resistance genes, often originating from promiscuous resistance plasmids and transposons. Phage vectors and phasmid vectors are also extensively used, and may have special relevance for the evolution of pathogenicity islands in bacterial pathogens. Thus, genetic engineering biotechnology has effectively opened up highways for horizontal gene transfer and recombination, where previously, there was only restricted access through narrow, tortuous footpaths. These gene transfer highways connect species in every Domain and Kingdom with the microbial populations via the universal mixing vessel, E. coli.

It is not easy to transfer genes successfully between species, there are barriers to horizontal gene transfers. That is why, apart from transposons which are promiscuous, such events were relatively rare in our evolutionary past. For example, analyses of 145 nonvertebrate globin gene sequences showed that there were probably two cases of horizontal gene transfer, one from the common ancestor of ciliates and the green algae to the ancestor of cyanobacteria and the other, from the ancestor of the yeasts to the ancestor of bacteria. Natural gene transfer vectors-viruses, plasmids and transposons-have probably always existed, especially in the microbial species; but they were to varying degrees host specific, so that the frequency of conjugal transfers are higher between the same species than with other species.

Naturally isolated vectors-even those that carry antibiotic resistance genes and have therefore already undergone evolution by horizontal gene transfer-have not exhibited the same wide range of transfer capabilities. Horizontal gene transfers have been directly demonstrated between bacteria in the marine environment, in the freshwater environment, and in the soil. Again, in all the experiments, horizontal gene transfers were mediated by specially constructed hybrid plasmid vectors, of the sort used in genetic engineering. Horizontal gene transfer occurs preferentially in interfaces between air and water and in the sediment, and especially under nutrient depletion conditions, thus refuting the claim that nutrient-rich media are necessary to support horizontal gene transfer.

Horizontal gene transfer of antibiotic resistances has even been demonstrated in wastewater treatment ponds, the effluent from which is increasingly being used for irrigation in developing countries. Frequencies of horizontal gene transfer may be greater under natural conditions than in the laboratory. Horizontal gene transfer is not limited to the external environment. It has been demonstrated between gut bacteria in mice and chickens and in the gut as well as urogenital and respiratory tract of human beings.

DNA can persist in the environment: Genes carried by vectors as naked DNA as well as chromosomal DNA can survive indefinitely, especially when adsorbed to solid particles in all environment; on surfaces of water, as sediment, in the soil, where they are efficiently taken up by other microbes. Although DNA is rapidly broken down in waste water, adsorption to solid particles in the sludge, which happens very quickly, will stabilize it and prolong its transforming capacity.

In addition, DNA is now found to persist for long periods in the laboratory clinging to many different surfaces. Transformation by the uptake of DNA is a major route of horizontal gene transfer in the environment. DNA is not only released when cells die, but are actively secreted by living cells, and even fragments of genes can have significant effects when transferred. As increasing amounts of recombinant DNA, especially of the 'naked' variety, are discharged into the environment, the risks associated with DNA transformation will rise exponentially.

DNA can resist digestion in the gut: Earlier studies supported the assumption that DNA is broken down into small fragments in the stomach and subsequently further degraded by nucleases in the digestive system. But more recently, Schubbert and coworkers, using the much more sensitive PCR (polymerase chain reaction) technique, were able to detect large fragments of the M13 phage DNA, which were fed to mice, in the faeces, the blood stream and within white blood cells. The persistence of DNA in the gastrointestinal tract suggests that transformation of gut bacteria may occur. This is very relevant to the spread of antibiotic resistance marker genes which are currently in nearly all commercially approved transgenic food crops.

Subsequent studies by the same group revealed that the DNA fed to mice was taken up, not only by white blood cells, but also by spleen and liver cells. The M13 phage DNA was reisolated from spleen cells and some were found to be attached to mouse DNA, suggesting that it had become integrated. This raises general questions on the ability of gene transfer vectors to invade mammalian cells directly, and indeed, that has been known to occur.

Mammalian cells Invasion by vector DNA and viral DNA: Bacterial viruses and baculovirus can also be taken up by mammalian cells. One main route for the uptake of DNA by mammalian cells is by phagocytosis of bacteria containing the vector DNA. But baculovirus can gain access by itself. It is so good at getting into human liver cells that it is being engineered as a vector for gene replacement therapy at the same time that it is being engineered in agriculture to be a more effective insecticide, which should raise serious safety concerns. An important route for horizontal gene transfer from bacteria to mammalian cells may be conjugation.

Bacterial conjugative plasmids can mobilize DNA transfer between bacteria and yeast. It was predicted that these plasmids raise the possibility that 'conjugation could occur between bacteria and animal cells and between bacteria and plant cells'. This prediction has been borne out. Recently, the prokaryote-eukaryote shuttle vectors pAT497 and pAT498 were both transferred from E. coli cells into mammalian cells by conjugation, with a helper plasmid from Shigella. Once inside cells, the DNA may integrate at random into the host cell genome, causing correspondingly random genetic disturbances. Of particular concerns in this regard are vector DNA engineered with oncogenes and other viral sequences which can cause cancers. Recombination between vector DNA and endogenous viruses can also generate new viral pathogens implicated in several types of animal cancers.

Direct documentation of the acquisition of antibiotic resistance: The antibiotic streptothricin was administered to pigs beginning in 1982. By 1983, plasmids encoding streptothricin resistance was found in the pig gut bacteria. This has spread to the gut bacteria of farmworkers and their family members by 1984, and to the general public and pathological strains of bacteria the following year. The antibiotic was withdrawn in 1990. Yet the prevalence of the resistance plasmid has remained high when monitored in 1993, confirming the ability of microbial populations to serve as stable reservoirs for replication, recombination and horizontal gene transfer, in the absence of 'selective pressure'.

In a direct test of persistence of streptomycin-resistance, researchers cultured many independent lines of a streptomycin-resistant mutant of E. coli in the absence of the antibiotic. They found that all retained the resistance after 180 generations. Furthermore, the lines have, in the mean time, accumulated compensatory mutations in other parts of the genome which increased their competitive ability relative to the wild-type. Again, this demonstrates the irrelevance of conventional explanations of the 'natural selection of random mutations' with regard to the evolution of antibiotic resistance.

Even if chemical inactivation methods are 100% effective, large amounts of recombinant DNA have already been, and continue to be routinely released into the environment in one form or another, either in the sewage or directly spread onto fields. Many kinds of dangerous recombinant DNA, containing pathogenic sequences, cancer-causing viruses, as well as antibiotic resistance genes can almost certainly transform bacteria in the environment and further recombine, spreading not only antibiotic and virulence genes in bacteria, but also generating new viral pathogens. There is an urgent need to re-assess the safety regulation, not only of deliberate releases, but also of contained use.

The antibiotics are currently used in treating TB, the top killer in the world to-day. This news item was put out by one of the EPA's own officials, who has been protesting on the internet against the agency's effective lack of biosafety oversight on genetically engineered products.

CHAPTER 6

OPPORTUNITIES IN AGRICULTURE

World food production has doubled since 1960 due to the scientific advances in crop breeding and farming techniques. Also productivity from agricultural land and water usage has tripled. Today, the peoples of the world farm an area about the size of South America; without the scientific advances of the past 30 years, farmland equivalent to the entire Western hemisphere would be required.

The world's population is expected to double by the year 2030 to 12 billion, and it is not clear whether current food production is keeping pace with population growth. The question is how best to feed billions of additional people without destroying much of the planet in the process. The disappearance of tropical rain forests, wetlands, and other vital habitats will accelerate unless agriculture somehow becomes more productive and less taxing to the environment.

Biotechnology can be employed to improve the quality of seed grains; increase protein levels in forage crops; and instill in crops resistance to disease, insects, and viruses, as well as tolerance for droughts, floods, and extreme temperatures. In addition, biotechnology can make foods healthier and more nutritious. For example, tomatoes and other fruits and vegetables containing increased levels of certain nutrients, such as vitamins

C and E and beta carotene, may help protect against risk of chronic diseases, such as some cancers and heart disease.

Although major genetic improvements have been made in crops, progress in conventional breeding programs has been slow. Moreover, most crops grown in the United States produce less than 50 percent of their genetic potential. These shortfalls in yield are due in large part to the inability of crops to tolerate or adapt to environmental stresses, pests, and disease.

For example, some of the world's highest yields of potatoes are in Idaho under irrigation, but in 1993 both quality and yield were reduced severely because of cold, wet weather and widespread frost damage during June. And some of the world's best bread wheats and malting barleys are produced in the north-central states, but in 1993 the disease Fusarium caused an estimated $1 billion in damage.

Major advances also have been made through conventional breeding and selection of livestock. Feed efficiency for poultry and swine has been increased by nearly 50 percent. Milk production per cow has more than doubled since 1955. Diseases such as hog cholera and pests such as screwworm have been eradicated. Yet major diseases of livestock still go uncontrolled and some still go undiagnosed or even unrecognized.

These and many other agricultural production hazards have defied traditional solutions such as classical breeding approaches. Biotechnology can compress the time frame required to translate fundamental discoveries into applications. With improved technology and knowledge about agricultural organisms, processes, and ecosystems, opportunities will emerge to produce new and improved agricultural products in an environmentally sound manner.

GENE SEQUENCING AND MAPPING

Continue mapping and sequencing of animal/plant/microbial genomes to elucidate gene function and regulation, and to facilitate the discovery of new genes as a prelude to gene modification. To locate desired genetic traits of organisms and effectively use molecular probes to identify DNA sequences, researchers need precise and relatively complete genetic maps. The combination of detailed maps and DNA sequences of genomes can provide a guide to plant and animal breeders for the development of new, efficient, and informed breeding strategies.

The Federally funded Human Genome Project is rapidly advancing gene mapping and sequencing technologies as well as data management techniques for analyzing genome data from humans and other organisms. These advances have the potential to makea important contributions to the genetic improvement of crops and animals. The agricultural research community is in an ideal position to apply the technical and scientific advances from the Human Genome Project to address the biology of organisms important to agriculture, aquaculture, and forestry.

Plant Genome

The U.S. agricultural research community has mounted a considerable effort to understand the structure, function, and regulation of genes in a wide range of crop and forestry species. There may never be resources sufficient to sequence completely the genomes of all agriculturally important plants, but there is a high degree of similarity among species. Therefore, it is cost-effective to select one representative plant for concerted sequencing of its entire genome.

The model experimental plant *Arabidopsis thaliana* offers the best chance for achieving a completely sequenced plant genome in the foreseeable future. *Arabidopsis* has the smallest known genome of any flowering plant, with 100 million base pairs and little repetition in its DNA. Detailed genetic and physical maps have been developed, and understanding of the plant's biology is growing rapidly. A complete understanding of the *Arabidopsis* genome would offer enormous potential for improving agriculture. In addition to studies on *Arabidopsis*, parallel studies will be required at some level with plants used to grow food and fiber.

Knowledge of a single gene can have broad impact. An example is a recent breakthrough in the identification and sequencing of a tomato gene that confers resistance to a bacterial disease. The sequencing of this gene has given scientists the first unambiguous glimpse of the resistance response of a plant to one of its pathogens. This discovery has opened the way to identification of disease resistance genes in other plants as well as created an avenue for investigating the molecular basis for disease resistance.

Genetics of Rhizosphere Microorganisms

Most farms grow only three or four crops, and some grow the same crops year after year in the same fields. These practices promote soil infestations by pathogens that damage or destroy roots. For most crops, soilborne plant pathogens generally are uncontrolled, except by limited crop rotations and some host plant resistance. Recent studies show that some soil microorganisms associated with the roots of crops carry and express genes for defense of roots against pathogens.

Little is known about these beneficial microorganisms, which represent an enormous untapped genetic resource for improving crop production and efficiency of fertilizer use. Identification and genetic studies of these microorganisms could enable their direct deployment or use of their genes in the development of disease-resistant crops. The agricultural research community has been quick to take advantage of findings from the Human Genome Project, such as the identification of the gene sequence coding for protein C. Protein C is needed for proper blood coagulation and can be used to treat blood disorders such as hemophilia.

Scientists have inserted the protein C gene into swine and developed animals that produce protein C in their milk. The use of swine instead of human blood donations for protein C production would avert the potential contamination of human medications with human serum viruses.

Also, production costs would be low because a small number of lactating sows would be able to meet the medical demand for protein C. An organized, coordinated national research program is in progress to enhance understanding of the structure, function, and regulation of genes in economically important animal species. To achieve maximum effectiveness and cost efficiency, this effort is being conducted in cooperation with other international genome mapping programs. One important focus is on traits controlled by multiple genes.

Most genetic traits that are economically significant are encoded and controlled by quantitative trait loci. In animals, these sets of genes control characteristics such as leanness of meat, meat tenderness, disease resistance, and, in dairy cattle, miik quality and quantity. Recent advances offer the possibility of using molecular markers to improve traits controlled by quantitative trait loci. In some cases, quantitative trait loci can be dissected into individual components, which then become amenable to manipulation. More research is needed to translate the advances made with single-gene traits into advances in understanding of quantitative traits.

Metabolic Pathways

Determine biochemical and genetic control mechanisms of metabolic pathways in animals, plants, and microbes that may lead to products with novel food, pharmaceutical, and industrial uses. In the past, new products were developed by exploiting natural materials. Antibiotics were derived from microbes, spices and perfumes from plants, and pharmaceutical agents from plants and other organisms. Plants also have served as renewable resources for energy, building materials, and industrial materials. Today, the tools of biotechnology offer new ways to exploit these biological resources to the maximum benefit of society.

The genetic manipulation of plant biochemical pathways can create new products for food and nonfood uses that will expand markets and increase the value of agricultural commodities. Metabolic studies, which examine the aggregate of the chemical processes carried out by an organism, are a logical focus of Federal support. The ability to add or delete specific genes and change the level of enzyme activity in transgenic plants has given scientists a new way of manipulating and exploring plant metabolism. The development and study of transgenic potatoes, for example, has made it possible to identify the key enzymatic steps in the starch biosynthetic pathway and has led to development of a new variety of potato that has more starch and less water.

Only a fraction of the estimated 100,000 compounds — including lignin, cellulose, terpenoids, and lipids — produced by biochemical pathways in higher plants have been characterized. The dearth of knowledge is even more extreme for lower plants, such as algae and mosses. There is a pressing need to understand the intricate mechanisms that control various pathways. Research on metabolic pathways and their biochemical and genetic regulation should be expanded, because genes of use to agriculture are likely to be found in diverse organisms. Plants may be the organisms of choice for production of specialty chemicals efficiently and in large volume.

Cell Wall Biomass

The cell walls of higher plants are the Earth's most abundant form of biomass. The structure and strength of wood and its industrial products are determined by the nature of the cell walls. Cell walls are the major component of forage crops; the chemical composition of the walls apparently affects feed digestibility.

Cell walls provide the mechanical stabilization of the plant body and influence the ultimate size and shape of the plant. Therefore, scientists must understand the biochemistry and physics of cell walls before they can employ biotechnology to modify the architecture of plants.

While progress has been made in understanding plant cell walls, many key questions remain. For example, not a single major enzyme involved in the synthesis of cell wall polysaccharides has been isolated. Information is needed about the organization of individual components within cell walls. And the regulation of the synthesis and assembly of cell wall materials in response to developmental and environmental signals is poorly understood.

Higher plants are very efficient sources of renewable organic materials. In particular, corn starch, which is both a useful raw material and an important human and animal food, is produced in large quantities at about the same cost per pound as is crude petroleum. Many genes involved in starch biosynthesis have been cloned recently and it should be possible to explore the development of new kinds of starches for use in nonfood applications. The potential use of starch as a co-polymer in biodegradable plastics is being explored.

Plant storage lipids (i.e., oils) can be produced in large quantities, and they represent a chemically versatile form of biomass. Enhancement of oil yield and composition in plants such as palm and rapeseed depends on increased understanding of the factors that regulate lipid biosynthesis. Progress has been made in identifying genes and enzymes involved in the early steps of fatty acid and lipid biosynthesis. To date, the most promising approach to manipulating oil yield and composition involves modifying the vast variety of fatty acids found as lipid constituents in various wild species. This is technically possible only if the basic biochemistry is elucidated.

Novel Products from Plants

Higher plants synthesize an enormous spectrum of chemical constituents. Many have known value as drugs (e.g., taxol), biomaterials (e.g., rubber), solvents, flavorings, fragrances, or coloring agents. Through development of transgenic plants, the purity of many plant-derived chemicals can be enhanced and the range of chemicals expanded. One example is the pending development of plants that produce hydroxylated fatty acids. These oils can be used in applications ranging from hydraulic fluids to nylon synthesis.

Similarly, the recent development of plants that accumulate biodegradable thermoplastics illustrates the potential for producing completely new compounds in plants. Many natural chemicals are important in plant defense mechanisms or in abiotic stress responses. Some chemicals confer important horticultural qualities to plants. Thorough studies have been conducted on the biosynthesis of certain classes of these compounds, such as flavonoids, cyanogenic glycosides, and certain alkaloids (e.g., nicotine).

Other classes of compounds, such as terpenoids, have received relatively little attention. Terpenoids are important for plant growth and can be used as essential oils, resin acids, and pigments. An efficient mechanism must be developed to obtain specific biochemical information about these classes of compounds, to complement the extensive ongoing efforts to identify them and elucidate their structures. Biotechnology offers promise for accelerating the characterization of these substances as well as facilitating their production for commercial use.

Extend understanding of the biochemical and molecular basis of growth and development including structural biology of plants and animals. Biotechnology offers opportunities to study a wide variety of genetically regulated growth and development processes. Alteration of these processes can help improve product quality and nutritional and economic benefits.

The genetic makeup of plants and animals can be altered by either insertion of new, useful genes or removal of unwanted ones. Discoveries made through biotechnological approaches are changing the way plants and animals are grown, boosting their value to growers, processors, and consumers alike.

The numerous potential benefits range from the development of specialty starch polymers for packaging and oils for use as industrial lubricants to modification of post-harvest traits such as ripening and senescence. The recent development of tomatoes with modified ripening characteristics may lead to extended post-harvest shelf life of fruits, vegetables, and flowers. Information about the biological basis of cold tolerance may result in the development of cold-hardy flowers and ornamentals now grown only in subtropical or tropical climates. The engineering of increased salt tolerance in plants may offset losses due to saline soils.

Scientists are gaining their first glimpse into the complex mechanisms of flowering. Most plants use environmental cues to initiate flowering so that individuals of the same species flower synchronously, thus maximizing the chances of successful outcrossing. Scientists have discovered molecules in plants called phytochromes that sense dark and light conditions and send signals that regulate plant activities such as sprouting, growth, and flowering.

There is still much to learn about how flowers develop. Investigations of genes controlling floral meristems — the tissue that forms flowers — promise to help explain the genetic mechanisms that manage the establishment and maintenance of this tissue. Ongoing research also is extending understanding of the physiological processes that take place during the transition from vegetative to reproductive development.

From apples to zucchini, most foods originate from the fertilized flowers of plants. In tomatoes, for example, several genes have been discovered that act within a flower's anthers, where pollen is produced. Promoters (regions of genes that act as on-off switches) also have been identified. Scientists now are pinpointing the overall molecular mechanisms that cue tomato pollen to fertilize the female part of a tomato flower.

Increased understanding of plant fertilization eventually could lead to incremental improvements in the yield and energy efficiency of crop production. Some plants have genes that render their anthers infertile. This male sterility allows production of higher yielding hybrid seeds, whereby male-sterile plants are fertilized with pollen from specially selected sources. Genetically engineered systems for controlling male fertility are used to produce hybrid seed of important crops such as oilseed rape, corn, and rice.

The technology also is applicable to tomato, lettuce, and a wide range of other crop plants. These advances eliminate the need for costly, labor-intensive hand or mechanical removal of anthers in making hybrid crosses and will have a major impact on the billion-dollar hybrid seed industry. In addition, the knowledge gained by studying anther development and fertilization at the molecular and genetic levels will reveal other approaches that can be used to produce novel varieties of hybrid crop plants.

Conventional breeding often is aimed at achieving certain visual traits (e.g., tall or short plants, slender leaves). Once the biological basis for these traits is understood, biotechnology could provide the tools to change plant architecture precisely and optimize plant form. For example, changes in leaf form and/or number could maximize photosynthetic capacity and increase production, and changes in root architecture could facilitate and maximize water and mineral capture and uptake.

Also, the tools of biotechnology could be exploited to design crops that thrive on soils contaminated with heavy metals, such as cadmium or mercury. Scientists are studying and beginning to redesign the genes responsible for the activity of phytochelatins, natural compounds in plants that bind these metals and detoxify them.

ENVIRONMENTAL INTERACTIONS

Elucidate the molecular basis of interactions of plants and animals with their physical and biological environments, as a basis for improving the organisms' health and well-being. Future gains in agricultural productivity and sustainability depend heavily on the use of biotechnology to improve the health and well-being of agriculturally important plants and animals.

Research on environmental interactions is emerging rapidly as a powerful means of increasing resistance to stress and disease, with consequent economic and environmental benefits. For example, basic research on the genetic control of biochemical responses to water stress in drought-hardy native plants can suggest how crop plants might be modified genetically for more consistent performance in years of drought.

New crop genotypes also must be resistant to an expanding diversity of pests and diseases. Basic research at the organismal and cellular levels, together with classical breeding techniques, already has paid significant dividends to farmers, consumers, and the environment. Today, U.S. wheat, corn, soybean, barley, and sorghum are grown without fungicides because resistant varieties have been bred. However, these and most other U.S. crops still are subject to major damage from insects and diseases caused by soilborne plant pathogens and insect-vectored viruses, for which there has been no useful source of resistance.

The survival of high-value horticultural crops depends largely on soil fumigants. Biotechnology also can help reduce the billions of dollars in agricultural losses caused by animal diseases. There are opportunities to boost the immune competency of animal hosts by genetic manipulation, develop biotechnology-derived regulators of immune function, and identify ways to prevent attachment of disease agents to host cells.

Factors affecting the disease- producing capacity of a bacterium or virus can be identified, and strategies can be devised to help the host withstand these pathogens. Through the use of techniques such as enzyme-linked immunosorbent assays, polymerase chain reaction (PCR), and monoclonal antibody-based systems, biotechnology can contribute to major advances in diagnostics.

Plant Pests and Diseases

Virologists have transferred virus genes — such as those for production of virus coat proteins — to plants, thereby conferring resistance to those viruses in otherwise susceptible hosts. Some genes also have been shown to limit virus replication and associated damage. Replication depends on the regulatory and metabolic processes of the host; these processes must be understood before scientists can engineer virus-resistant plant varieties. Another possible way to increase plant resistance to viral infections would be to interfere with the spread of the virus, but this approach would require an understanding of the biochemical mechanisms of spread and transmission.

All crop plants are susceptible to bacterial diseases, which have been difficult to control through plant breeding. Bactericides are not a complete solution, either, because bacteria quickly evolve resistance to them. The bacterial disease fire blight is destructive to pears, apples, quince, and some ornamental plants. This disease has plagued fruit producers for over three centuries, but its mechanism of attack only recently has been revealed using the tools of molecular biology. This knowledge is expected to provide ways to combat this disease.

Nearly all agriculturally important fruit, vegetable, and grain varieties acquire fungal diseases that can cost growers billions of dollars annually. Information is accumulating rapidly on the molecular basis of plant responses to infection by these pathogens. In many cases, the difference between resistance and susceptibility is the rate of response; if rapid, then the plant is resistant.

Scientists have found that deliberately disarmed fungal pathogens can trigger the necessary response in susceptible (i.e., slow responding) plants before the virulent pathogen arrives. This approach has opened a new avenue in disease control: the use of "pathogen derived" biocontrol agents to induce resistance in otherwise susceptible plants. This technique is only beginning to emerge as a way to control heretofore unmanageable plant diseases.

Control of Insects

Insect pests cause major crop damage. Examples include *Phylloxera*, which has caused devastation in European and Californian vineyards, and *Psylla*, which has ravaged the apple and pear industry of the eastern United States. In this century, the United States has imported and released approximately 800 natural enemies of insect pests; about 40 percent are providing some level of biological control. However, this approach to insect pest management has its limits. Further progress will require increased knowledge of interactions between pests and their natural enemies, and, if necessary, appropriate genetic modifications.

The chemicals emitted by plants in response to feeding by insect pests serve as an attractant to the natural enemies of some insect pests. Corn leaves attacked by beet armyworm caterpillars release volatile compounds that attract wasps that are parasites of these armyworms. Moreover, these parasitic wasps can recognize and fly toward the source of volatile compounds produced in response to feeding by their caterpillar hosts; the wasps ignore the odors released from leaves damaged mechanically, such as by mowing.

Expanded knowledge of the "information molecules" used by beneficial insects to find their prey on crops could lead to development of crops capable of producing stronger signals, to attract higher populations of beneficial insects. To exploit fully this approach to biological control, more research is needed on the biochemistry and molecular biology of insect pests and their natural enemies. Due to insects' evolving resistance to pesticides, and the removal of many of these chemicals from the market, there are few effective means of controlling sucking insects, such as whiteflies, aphids, and leafhoppers.

These same insects are among the most important vectors (carriers) of viruses, rickettsia-like bacteria, and mycoplasma-like organisms responsible for major plant diseases. New synthetic chemical compounds to be on the market in the next few years primarily target chewing insects. Biological methods to control sucking insects also are advancing at the molecular level. An example is the enzymatic preparation of synthetic pyrethroids.

The salivary glands of aphids, whiteflies, and leafhoppers contain bacterial symbionts that provide essential amino acids to their insect hosts. Only a few of these symbionts have been characterized. Research is needed on the basic biology, molecular biology, and production of these bacterial symbionts. If the bacterial symbionts could be disrupted in some way, either by manipulating the genes of the symbionts or by engineering an antimicrobial agent into a plant, then an innovative control mechanism could be developed.

Beneficial Microorganisms

Plants selectively aid the growth of specific types of beneficial microorganisms. Some microorganisms, for instance, have been shown to provide growth factors for plants and protect plants against insect attack and infection. Information about the molecular signals and genetic controls of these associations can lead to the use of these microorganisms to improve the performance — and especially the consistency of performance — of important food and fiber crops, and even to enhance the effectiveness of the microbes themselves. As an example, legumes such as soybeans rely on microorganismsliving inside their roots to "

Work is underway to introduce new genes into animals in order to impart resistance to disease and to identify host cell traits that can be used to enhance resistance to disease agents. For instance, a specific receptor on pig intestinal cells is required for *E. coli* to bind to the cell. Strains that do not carry the receptor — or have the receptor deleted through the tools of biotechnology — are less susceptible to the diarrheal disease caused by the bacterium.

Other research on enteric rotavirus infections in pigs has identified specific cell membrane components required for attachment of the rotavirus to an intestinal cell. A complementary molecule can be constructed using biotechnological methods to saturate and block this receptor site. Porcine stress syndrome is an inherited condition in swine that can lead to meat that is of poor quality or unusable.

This condition causes significant losses in the swine industry. At the same time, animals susceptible to this genetic disease have highly favorable muscling with low fat content. A candidate gene for this disease has been identified and research now is needed to develop a DNA probe for the trait, as a means to identify susceptible animals and eliminate them from breeding programs.

ENHANCE FOOD SAFETY

Enhance food safety assurance methodologies, such as rapid tests for identifying chemical and biological contaminants in food and water. Food- and waterborne illnesses caused by microorganisms pathogenic to humans have been estimated to affect more than 80 million Americans and cost the U.S. economy over $40 billion annually. For example, some 9,000 Americans die each year from food-borne illnesses caused by microorganisms such as *E. coli* in meat and *Salmonella* in poultry. At present, virtually all food inspection is visual. Rapid, accurate, non-invasive tools of biotechnology can help improve the detection and control of food-borne human pathogens as well as chemical contaminants. Federal support is needed in this area to assure that these tools are developed in a timely manner.

DNA Probes

DNA probe kits have been developed for *Salmonella*, *Listeria*, *E.coli* 0157:H7, and *Staphylococcus aureus*. Compared to traditional culture-plating methods, these new diagnostic kits offer greater precision, shorter turnaround times, and reduced need for highly trained personnel. DNA diagnostic techniques for Norwalk viruses also are commercially available. Culture techniques have not been successful in detecting these leading causes of gastroenteritis, making the availability of genetically based techniques critical to the detection and identification of the Norwalk viruses.

In addition to detecting food contaminants, DNA probes and other tools of biotechnology can help reduce levels of naturally occurring toxicants in foods. DNA probes can be exploited in research and plant breeding to isolate genes associated with the biosynthesis of major toxicants, facilitate understanding of the genetic regulatory mechanisms for toxicants in plants, and develop lines of plants with reduced levels of toxicants.

Mycotoxins in food are a periodic threat to food safety. DNA probes could help detect the presence of mycotoxin-producing fungi that grow under certain conditions in plant materials such as improperly dried corn and peanuts. DNA probes could also be used to learn more about the sources of fungal contamination in the environment and as a means to develop management strategies under field conditions.

Biosensors

The development of biosensors offers great promise for improving food processing, analysis, and safety assurance. The highly specific actions of biological molecules can be exploited for use in biosensors that can measure the concentration of specific components in complex mixtures. Enzymes, antibodies, and microbial cells can be immobilized on solid surfaces, and the specific reactions they mediate can be detected by various physical and chemical means. Biosensors are commercially available to detect a variety of sugars, alcohols, esters, peptides, amino acids, cell types, and antibiotics.

Development of tailor-made membranes capable of separating molecules based on size, electrical charge, or solubility will accelerate biosensor development as well as the exploitation of biomimetic systems. Miniaturization and mass production of biosensors could increase their availability and decrease their unit cost. Technologies such as microlithography, ultrathin membranes, and molecular self-assembly have the potential to facilitate the miniaturization of molecular and cellular processes for enrichment, detection, and analysis of chemical and microbiological contaminants in food.

Advances in the semiconductor industry have made it possible to combine chemical and biological components and integrated circuits in miniaturized systems. Biosensors can be inserted directly into food processing streams to obtain on-line, real-time measurements of important food processing parameters. Miniature biosensors also could be incorporated into food packages to monitor temperature stress, microbial contamination, or remaining shelf life, and to provide a visual indicator to consumers of product state at the time of purchase.

Food-Borne Microbes

An appropriately designed long-term molecular biology study could characterize and correlate food-associated microbiological isolates in existing culture collections. Food-related isolates from these collections could be used to screen foods for microbiological contamination. Foods identified in this way then would serve as focal points for regulatory agencies and critical systems aimed at improving control over growing, shipping, processing, distribution, and other steps during which these foods may become contaminated.

CHAPTER 7

GENETIC ENGINEERING IN AGRICULTURE

Plant cell tissue culture is being used to produce plant-derived products. The first tissue culture product commercialised was Shikonin, a dye and pharmaceutical manufactured by Mitsui Ltd. in Japan, in 1983. Ginseng, one of the most valuable plant products, can be harvested only after 5-6 years of plant growth, and the plant is destroyed. Large yields have been produced by culture of chlorophyll-accumulating aggregates of ginseng cells. These two examples did not use genetic engineering techniques, which will expand the range enormously.

The technology is still at an early stage, and many pharmaceutical products can be produced. The advantage of using plant tissue culture is that factory control makes the supply independent of seasonal plant yields or imports. Transgenic plants are also being used to produce industrial products. Very recently two American biotechnology companies have begun to use plants to produce melanin, the natural pigment that darkens skin. This will be used in new sunscreen lotions.

There have also been pharmaceutical peptides produced in oilseed rape plants. Some of these proteins could be economically produced in the seeds of plants. There is much research interest in the production of thaumuatin, one of the sweetest known substances. It is normally extracted from a West African plant, however, the gene has been isolated.

For example, in New Zealand the gene has been transferred to potatoes, with the idea of using potato as a source of supply. Thaumatin is already used as a food additive to increase weight gain in pigs by making them eat more of the nicer tasting sweet food. If thaumatin is included in the crops directly it will be simpler. Also, the worldwide market for human low calorie sweeteners is over US$ 3 billion annually, so there is much possibility for entering this market with a lower cost product.

Human serum albumin has been produced in potato and tobacco plants. Animal antibodies have been produced in transgenic plants, and this could bring the cost of producing monoclonal antibodies down by a factor of 10,000. Animal antibodies could also be used to protect plants against viral diseases, or the antibodies could be used to scavenge small organic pollutants such as toxins, from the environment. Obviously, the potential range of proteins is enormous.

Bacteria can be used to produce polymers that can be processed into polypropylene-like plastic. Biopolymers can be made using the precise enzymatic control that is not possible to use with synthetic polymers, with the advantage of biodegradibility to avoid pollution problems. Bacteria can also be made to produce the raw material for biodegradable plastic bags. The genes for polymer production may be put into foodcrops, such as potato tubers. This would also avoid using nonrenewable and energy intensive production techniques.

PLANT BREEDING

For millenium plants and animals have been selectively bred to develop varieties that are more productive, or suitable for human use. The welfare of humanity is inextricably bound up with efficient agriculture. Plant breeding as a science began in the 19th century after the discoveries of how plant traits are inherited. Our modern varieties originated from gene transfers within crop species, by selective breeding. There are, however, some major exceptions.

For example, about 5,000 years ago wheat was created, when the three genomes of Triticum monococcum, Triticum tauschii, and a species of Aegilops came to be combined. The definition of a species rests on the concept of genetic isolation but sexual exchange of genes between species can and does occur in nature without human intervention.

Often, the crop species does not contain sufficient genetic diversity to allow the desired improvements, hence the search for diversity has led plant breeders to use new genetic technology. The same concern has led animal breeders to use the same technology, as well as a desire to increase the desirable traits beyond those obtainable by normal breeding. In both cases, the aim is to arrive at a breeding population consistently expressing the desired trait(s).

One of the main weaknesses of conventional plant breeding is its dependency upon sexual crosses and thus to genes that exist only in one species. There are dozens of examples of agriculturally important genes and traits transferred to crop plants by interspecific or intergenic hybridisation and they have been reviewed by several authors. Recombinant DNA technology allows the transcendence of inter-species barriers and makes very novel genetic combinations possible. The first transgenic plants were created in 1983.

One of the most popular methods of gene transfer is the use of the soil bacterium Agrobacterium tumefaciens, which can transfer genes to many plants at wound sites. However, it works mainly on the dicotyledonous plants which excludes many crop plants, such as cereals. Direct DNA transfer can be used to transfer genes to protoplasts from which plants can be regenerated. About 150 species of plants have already been regenerated from protoplasts, so the potential application of the technique is already very large.

Among the techniques for gene transfer another common one is "biolistics", the use of particle guns to shoot DNA into cells. Some techniques use tungsten particles, or gold beads.

There may also be advantages of up to a 40% reduction in time for crop production via some biolistic based approaches over using Agrobacterium. This approach has recently been used to transform corn, and for the first time to produce genetically-modified fertile corn.

The gene for resistance to the herbicide bialaphos has been inserted and the corn are resistant. Rice and corn are the only major crops that have been genetically modified to produce fertile transgenic offspring. Further improvements should be expected and developed, and the final proof of success must wait until field studies are completed. Microinjection has also got potential. Gene transfer technology has advanced at a far faster pace than our understanding of plant biotechnology and the factors which are important within the plant in determining other useful agronomic traits.

Because of this, attention has been focused largely on characters which might be determined by single genes. The main focus of most biotechnology programmes is to produce new cultivars with improved pest and disease resistance to promote more environmentally acceptable alternatives for food production. Yield is no longer the only goal, improving the quality and marketing appeal, using genetically engineered pest and disease resistance to produce healthier products, are goals. Traditional methods of cross-breeding are not only limited, but slow and costly. Transgenic plants are also being used to study gene expression and control, which will be useful for further practical applications.

Plant Disease Resistance

Natural disease resistance is complicated. Plant breeders have long sought to increase the disease resistance of crops through selection of resistant varieties and by hybridising crops with wild relatives. About one third of total crop losses are directly attributable to plant disease. Molecular techniques, such as insertion of antiviral or antibacterial genes from other species into plants, and cellular methods to allow rapid screening for the desired phenotype, have led to more rapid progress.

Plant viruses are not well understood, but opportunities for improved disease resistance exist as the plant-pathogen relationship is easier to modify. Viruses cause serious diseases in many crops. The genetic basis of viral resistance in plants is narrow, so resistance breaking strains of virus frequently appear. Isolating the plant's own resistance genes to combat disease is not practical until they have been isolated.

The function of such genes depends on complex factors, such as the right genomic background. However, they could be used as good starting materials for protein engineering. Good viral disease control has been obtained using three different approaches: Cross protection occurs when plants are deliberately inoculated with a mild strain of virus. The plants are then resistant to infection with normal virulent strains of the same virus.

Cross protection is probably due to the presence of the coat protein of the protecting strain. Coat protein genes of several viruses have been inserted into transgenic plants to provide protection. The expression of a single coat protein gene can protect a plant against several different viruses. This heterologous protection is important, as it reduces the number of genes required for multiple viral resistance.

This has protected crops from infections of alfalfa mosaic virus, cucumber mosaic virus, potato viruses X and Y, soybean mosaic virus, tobacco mosaic virus, tobacco rattle virus, and others. Importantly, it has been found that transgenic plants that express the coat protein genes of tobacco mosaic virus or alfalfa mosaic virus also have some protection against other viruses. The mechanism of this protection is unknown.

Insertion of Satellite viruses (which are unable to replicate themselves) into the plants' genome to provide protection has been used for cucumber mosaic virus and tobacco ringspot virus. Antisense RNA the translation of a specific mRNA can be inhibited if the plant contains a complementary antisense RNA, which will form a double-stranded RNA molecule with part of the messenger mRNA, preventing translation of the protein, and thus protecting the plant.

Tobacco plants that have been made resistant to tobacco mosaic virus (TMV) infection, are of great commercial importance. Cotton plants have successfully expressed genetically transferred marker enzymes, and commercially important genes are being tested. Douglas fir has also expressed marker genes, and the seedlings were micropropagated from shoots.

The importance of forestry trees is very high, and disease resistance genes are perhaps the most immediate targets. There have also been genes inserted into tobacco, cotton, corn and soybeans to make them resistant to crown gall disease. Potato is one of the most important stable food crops worldwide. Because it is a tetraploid it is laborious to use traditional breeding to improve varieties.

Viral diseases are of major importance: potato virus X can cause yield depressions of over 10%, and potato virus Y can decrease yields by up to 80%. Thus virus resistance holds a major key to potato crop improvement. Potato plants containing resistance genes to these viruses, have been tested in many countries for several years. The inserted genes may not affect other cultivar characteristics. Current work is aimed at obtaining simultaneous resistance to more of the major viruses.

One team in Monsanto has generated resistance to both virus X and Y in a commercial potato variety, Russet Burbank. Bacteria can be used for disease resistance. In 1988 a bacterial fungicide called "Dagger G" was introduced for the control of cotton diseases caused by Rhizoctonia and Pythium. It was given approval in the USA by the EPA after only five months. This is in contrast to the common ten year period required for approval of chemicals. Dagger G should have fewer negative side effects than chemicals and will probably be improved by genetic modification.

Pest-Resistant Plants

The current pesticide market in the United States is worth US$ 3 billion annually. There are many problems associated with pesticide use, including pest resistance to chemicals and negative environmental effects. There are also insecticides being developed that are of plant origin.

Biocontrol

Biological control has a long history and is becoming more important. In 1889 Vedalia beetles were introduced to California to control cottony cushion scale in citrus orchards. They are still being used. In recent years imported parasitic wasps have been used to control alfalfa weevils. Another example is the introduction of musk thistle weevil which is a long term alternative to chemical herbicide for the control of musk thistle. It has been well used in North America. Another is the Ecogen product "Collego", for controlling northern gointretch, a costly weed for rice farmers. It is a dry formulation of a naturally occurring fungus. There are other similar products on the market, and more are being developed.

Crop rotation and tillage that disrupt pest life cycles are at the first level of control. At the second level is the introduction of self-sustaining control agents. Expansion of the marketing of biocontrol agents by major companies has been slow, as they will lose much profit from the sale of chemicals. There has been more introduction of repeated application of biological pest control, such as bacteria or fungi-based pesticides. Pests take longer to develop resistance to biocontrols, and although the results may take longer to see then with chemicals, the management can be long term. They have the additional advantage of specificity, not seen with insecticidal chemicals.

Insecticidal proteins

Plants expressing the insecticidal protein of a bacterium, Bacillus thuringiensis are pest resistant. Insect pests will die if they eat the plants. Larvae of moths and butterflies can be selectively killed by different insecticidal proteins. Many companies have put this gene into crop plants including corn, cotton, soybean, tobacco and tomato and this protected the plants from insect larvae. Some crops such as cotton may normally have over ten applications of insecticide over the growing season.

The Belgium company Plant Genetic Systems which has been testing plants expressing insecticidal proteins since 1986 has found the resistance to be very good without affecting yield. Like several US companies, including Monsanto, they will be producing several commercial varieties soon. The control of caterpillar pests with plants expressing this insecticidal gene offers several advantages. Control is independent of the weather, and in conditions which would be unsuitable for spraying chemicals or bacteria, the crop is still protected.

All parts of the plant are protected, such as the roots, or new growth previously susceptible between sprayings. The pests are affected as soon as they begin to feed. Broad spectrum insecticides kill all insects, which includes spiders and beetles which are useful predators. The B. thuringiensis endotoxin kills only killing leaf-eating species. Different insecticidal proteins have been expressed to kill larvae of Lepidoptera and Coleoptera (beetles). There are different proteins produced by different strains with varying specificity. Being proteins, they are biodegradable, and can be much cheaper to develop, and to obtain environmental release approval for use.

To develop and register a new chemical insecticide costs about US$ 25 million, but to develop a new plant variety costs about US$ 1 million. However, the Environmental Protection Agency (EPA) regulations may require that a new crop variety needs to be registered as a pesticide in the USA, which would add about US$ 10 million to the cost. The situation remains uncertain, and test crops are under review.

Digestive inhibitors

An alternative way to control herbivorous insect pests is by introducing the gene for digestive protease inhibitors into the plants, so the animals cannot digest food. The expression of these plant genes, which are thought to be a defensive response to insect attack, can be enhanced. Wounded plants produce a factor which induces the synthesis of protease inhibitors specific against insect and microbial proteases.

They have an effect on a wide range of insects and are known not to be harmful to humans. The big environmental advantage is that only insects that eat the plants are affected. Commercial seed are soon expected to be available. The first field tests in Belgium were in 1985. In New Zealand, researchers are trying to identify the best plant protease inhibitors for insertion into clover or pasture crops, there is also work using insecticidal proteins to improve pest resistance in clover.

Insecticidal microorganisms

There has also been work on the development of insecticidal microorganisms to be sprayed onto plants. The current application costs of spraying microorganisms containing a toxin gene are similar to the costs of applying chemicals, but with the significant environmental advantages. These need continual application, but may not require additional regulatory approval for human consumption, as they will need to be if they contain novel genes. Losses to crops also occur during storage after harvest. It is possible that increased levels of antifeedant could be added to plants to reduce such losses.

HERBICIDE TOLERANCE AND WEED CONTROL

The use of agrochemicals is still expanding. Weeds reduce crop productivity by at least 12%. Overall worldwide sales of herbicides are worth over $US 5 billion annually, and are double the market for insecticides, and fungicides. Even in developing countries where labour for weed control is very cheap, large quantities of herbicides are used. The effects of this large-scale chemical use are still unknown, and many effects are indirect.

The function of soil ecosystems is poorly understood, and the major biological cycles of organic matter breakdown and nitrification are the most sensitive to these chemicals. Genes that give plants tolerance to herbicides have been isolated and incorporated into some plants. Work has concentrated on herbicides that are more environmentally friendly than those commonly used. The gene from bacteria that confers resistance to the herbicide glyphosate (Roundup) has been expressed in higher plant chloroplasts.

Since 1987 tomato plants expressing the gene, have been made. The gene is under patent by an American company Calgene, and is called "GlyphoTol". It has also been tested in cotton and has been transferred to many other crops. Resistance to sulfonyl-urea compounds, the active ingredients in Glean and Oust herbicides, has been obtained by the introduction of a mutant acetolactate synthase gene. Many researchers have transferred the gene to a wide range of plants.

Plants resistant to phosphinothricin (Basta), have also been made by Plant Genetic Systems. Potato and tomato plants that can grow with concentrations of herbicide ten times higher than normal herbicide application level have been grown. The field can be sprayed to kill weeds without affecting the crops. The results indicate there is no yield penalty. Many field trials have been conducted. Resistance to other herbicides such as atrazine (AAtrex) is also being developed.

There are different ways to alter herbicide tolerance and a recent comprehensive review by Schulz et al. summarises these. Sensitive enzymes can be over-expressed, by increasing the copy number and/or the expression of the gene. It therefore requires more herbicide to kill the plant. Site-directed mutagenesis can be used to alter the herbicide binding to enzymes. The herbicide uptake can be reduced by altering transport systems or the morphological changes such as increasing the number of cuticular wax layers obstructing penetration of the herbicide.

Enzymes that degrade the herbicide can be expressed, and this approach has been effective against bromoxynil, phosphinothricin and 2,4-D. Research has mainly been conducted on those herbicides with properties such as high unit activity, low toxicity, low soil mobility, and rapid biodegradation and with broad spectrum activity against various weeds. The development of crop plants that are more tolerant to such herbicides should prove more effective, less costly and more environmentally attractive weed control. The commercial strategy for chemical companies is to gain increasing market share through a shift in herbicide use, not to increase overall use.

Herbicide tolerant plants will have the positive impact of reducing the overall herbicide use and will also lead to substitution by more effective and environmentally acceptable products. The first herbicide resistant crop to be released for use was atrazine-resistant Canola. Imidazolinone resistant corn and sulfonylurea resistant soybeans and potatoes are under evaluation. The technology is widely applicable, so once the food is judged to be safe, a large range of crops may be available.

It is possible to use phenotypic selection to attempt to develop herbicide-tolerant crop cultivars. Cellular selection has often been used, but the whole plant response can be different to the cellular response. Phenotypic recurrent selection has been used to select for 2,4-D tolerance in red clover. The levels of 2,4-D tolerance were increased by 35% over four cycles. This is an alternative to gene insertion. The advantage is that the gene is found "naturally" in that species.

The Soybean variety, Tracy M, was bred by classical means for resistance to the herbicide Metribuzin. However, this sort of selection method is not necessarily intrinsically safer for animal or human consumption. We need to look at genetic engineering as an extension of traditional crop breeding goals. All alternatives are potentially useful. There are several advantages of herbicide tolerant plants. The obvious use is in removing weeds from crops. It also allows the maintenance of genetic purity during seed multiplication of new cultivars.

It could allow chemical thinning of crops after the mixing of parent and resistant seeds. It can also be linked to other characters, as a selection method. There is no need to apply herbicides until weed infestation reaches an intolerable level, therefore less herbicide is used. This reduces soil erosion. A single safe herbicide can be used instead of a mixture of herbicides, reducing the chance of weeds becoming resistant to several herbicides.

Varieties in Crop

The most obvious improvement accomplished by traditional breeding is increased yield. In the USA from 1930-1975 the average yield per unit land increased by the following percentages: soybeans 70%, wheat 115%, corn 320%, and grain sorghum 358%. This increase was accomplished by small increments, but genetic engineering techniques have the potential to rapidly increase yield, as they complement the traditional technology.

In the USA the average annual increase in corn yield is 1%, however, in parts of southeast Asia where more people live the rice yield is no longer increasing. This presents problems that will require new approaches to improving yield. During the Green Revolution the average rice yield increased 4% annually, but principally by converting stalk to seed.

There is much potential to increase yield, for example by improving the efficiency of photosynthesis so that more carbon is fixed into plant material. There are efforts to alter the cell membranes of plants, as well as basic enzymes involved in photosynthesis, to increase the efficiency of plant growth. Any approach must also involve the associated good farming practise.

Environmental extremes

Plants may be able to be more resistant to drought, flooding, salinity or sensitivity to heavy metals, so that they can be grown in areas of the earth currently beyond the tolerance range of species, or even those areas unable to be used for agriculture at all. About 30% of the world's land area has major plant stress conditions, including insufficient soil nutrients or water, or toxic excesses of minerals and salts. To exploit other environments, tolerance to low temperature is also important.

The antifreeze gene from an arctic fish has been transferred to soybean, with the goal of creating plants tolerant to low temperature. There is research by a number of groups on the development of aluminium resistance in plants. Aluminium toxicity is a problem in low pH soils, where it may reduce plant growth. By making plants tolerant, they will grow better in such soils. Pine trees are being made more drought resistant and suited to warmer weather, because of the expected climatic changes due to global warming expected in North America in 30 or 40 years when the trees mature.

Due to the long reproductive cycle, and the need to wait 20-30 years before mature traits can be evaluated, we are now using only the second and third generations of genetically improved trees. The long juvenile periods, large size and high natural heterozygosity limit the application of conventional breeding techniques, so genetic engineering is more applicable to tree improvement than to herbaceous agronomic crops. The traits that will be targeted include climatic adaptation, fusiform rust resistance, and herbicide resistance to allow better plantation establishment. There are other long term targets such as nitrogen fixation, lignin biosynthesis, cellulose biosynthesis, photosynthetic efficiency, cytoplasmic male sterility, and apical dominance.

Nitrogen fixation

A major long term project for crop improvement is to characterise, then transfer, the genes for nitrogen fixation into plants to enable them to fix atmospheric nitrogen to save using nitrogen fertilisers. However, the nitrogen fixing pathway involves 17 different genes and regulation is important. The transfer of the unit of 17 linked genes has been done from Rhizobium to Aztobacter strains. The strains could then fix atmospheric nitrogen. To fix nitrogen for plants, there is also the requirement to have the genes for a symbiotic relationship between the legume and the symbiont.

The importance of this technology is highlighted by the growing pollution of ground water by nitrogenous fertilisers. Expensive biological and mechanical filtering to remove nitrates from drinking water is the current "solution". The only success has been in improving the nitrogen fixing ability of some Rhizobium strains. It is a much more distant goal to insert the genes into plants in a way in which they could be used. In terms of economics, the market for a nitrogen-fixing crop plant would probably be small in the developed world. The advantages for the third world would be greater, saving the costs of importing chemical fertilisers.

Improved nutritional qualities

The food content of seeds, and plant products can be altered to improve their nutritional qualities. New techniques are continually being developed. One such technology involves using antisense RNA sequences to bind to the mRNAs of undesired proteins. One application was the reduction of the concentration of an enzyme (polygalacturonase) which is produced by ripening tomatos causing softening of the tomato. The concentration of this enzyme was reduced by 99%, so the fruit stay firm.

These tomatoes have been developed to improve shelf life and taste since growers can leave the tomatoes on the plant longer. These tomatos are being patented by Calgene, and are currently under consideration for FDA approval for human consumption. There is work in New Zealand on the control of ripening and softening in apples. The goals are to improve storage life, as well as quality.

Entirely synthetic proteins have been designed to supplement the overall production of essential amino acids in potatoes. The appropriate DNA sequence can be designed and synthesised in the laboratory, then inserted into the crop. Insertion of an artificial gene coding for a protein rich in essential amino acids can reduce from 1.8 to 0.8 kg the amount of potato one must eat each day to get all the daily protein intake. In CSIRO, Australia and DSIR, New Zealand, pea albumin (which has many sulphur amino acids) is being inserted into white clover, so that it will lead to better wool growth in the sheep that feed on the improved clover. Improving the amount of methionine-rich protein in plants is important for other animal based systems also.

Widening consumer choice

There is work at the DSIR, in New Zealand, on introducing new characteristics to increase the number of varieties of kiwifruit. These include altering internal colour, texture, seed content and shape, and making smooth skin. There are also goals involving nutritional characteristics such as altering sweetness and increasing vitamin C. The aim is to introduce new varieties for consumer choice. There is also work on improving the flavour of onions. This type of food product requires examination before being used for human consumption.

Ornamental plants

There have also been advances in the breeding of ornamental plants. Exports of ornamentals from the Netherlands alone were worth US$ 200 million in 1987. Ornamental plants were one of the first concerns of early plant hybridisers. However, they have used less sophisticated methods than those for breeding field crops. With the development of genetic engineering, and given the economic importance, there is renewed interest. The choice of flower colour can be extended, as novelty is added, such as rare blues or purples. Flower colour manipulation was first reported in Germany in 1987 with petunias.

There have been some changes made using antisense methods. More long term objectives will be altering flower morphology, and improving vase life. Productivity will also be improved, as with other plants, by disease and pest resistance. A Netherlands company, Florigene, is producing genetically engineered roses, carnations, chrysanthemums and gerberas, with new colours. Unlike foodstuffs, there will not need to be proof that they are safe for human consumption.

Other plant properties

Other desirable properties in crops include the development of plants that require minimum attention, a low number of cultivation operations, and developing crops that can be mechanically harvested. Orchard trees need to be bred that conform to a given space, so control of branching is another goal.

The crops should all be ripe for harvesting at the same time. The characteristics of crops sent to further food processing steps after harvesting can also be altered to suit the machinery, e.g. the milling and baking qualities of wheat can be improved. A basic improvement is to increase the ratio of edible to nonedible parts in a plant, as was the basis of the changes to rice during the green revolution two decades ago.

MICROORGANISMS

There have been, and are many future possible uses of microorganisms in the environment, and this range has been greatly expanded by genetic engineering. The novelty of biotechnology is its ability to exploit the universality of the genetic code to combine, in a single organism, major adaptive traits developed by very different organisms. Some of the current uses are summarised below.

— Bacteria can be used as pesticides, for example to carry the Bacillus thuringiensis insecticidal proteins.

— Bacteria (Pseudomonas syringae) that have reduced ice-nucleating ability have been sprayed onto plants to reduce frost damage. Conversely, iIce-nucleating bacteria and the ice-nucleating protein from P. syringae have been successfully tested for use in artificial snow-making.

— Viruses (Baculovirus) can also be used as pesticides.

— There are many possible uses of viral vaccines, against human diseases such as Hepatitis A, B, and C polio, rabies, malaria and against animal diseases such as foot and mouth disease, sheep foot rot, or tapeworms.

— Plant symbionts can be introduced, such as nitrogen fixing bacteria, which reduce the need for fertilisers. Mycorrhizal fungi can be used to increase plant growth rates by improving the efficiency of root uptake of nutrients.

— Bacteria can be used to degrade toxic compounds, such as heavy metals, organic compounds, phosphorus, ammonia or other pollutants.

The first major use of bacterial degradation of an oil spill followed the Exon oil spill in Prince William Sound, Alaska in 1988. In this case, genetically modified bacteria were not added, but fertiliser to allow the low number of naturally occuring bacteria present in any soil that can degrade hydrocarbons, to multiply. The fertiliser is called Inipol. The technique is only applicable to beach areas, as the fertiliser will not cling to rock surfaces. The treated areas were dramatically improved within fifteen days, and even the underlying soil at a depth of one foot was degraded within 40-50 days. This was particularly important in the low temperature arctic environment where oil degrades very slowly.

Chapter 8

Impacts of Genetically Modified Crops

Research on GM crops for uses in agriculture started in the eighties but sales of first commodity seeds began only in the mid-nineties. The first significant sowings of GM crops took place in 1996 and almost exclusively in the US. Since 1996, the areas have increased dramatically to reach 41.5 Mio hectares in 1999. Adoption of transgenic crops is progressing at a much faster pace than has been the case for other innovations in plant varieties, e.g. hybrids. First indications on 2000 sowings of GM crops could be found in various sources, but they point to divergent directions. DG AGRI expects the GM area for 2000 to plateau just above 42 Mio ha.

As shown in table 1 most of the GM crops are grown on the American continent. In 1999, the US had by far the most important area of GM crops, around 70% of the total GM area worldwide, followed by Argentina and Canada. In China, the GM area ranks between 1 and 1.3 Mio ha, depending on the sources. This would represent about 3% of the 1999 world GM area. On the European continent in 1999, Spain ranked first with around 10000 ha followed by Romania with 2000 ha and France, Portugal and Ukraine at just 1000 ha.

Table 1: Development of GM area by country

Mio ha	1996	1997	1998	1999	1999 in %
USA	1,45	7,16	20,83	28,64	69,1%
Argentina	0,05	1,47	3,53	5,81	14,0%
Canada	0,11	1,68	2,75	4,01	9,7%
China	1,00	1,00	1,10	1,30	3,1%
Brazil	0,00	0,00	0,00	1,18	2,8%
Australia	0,00	0,20	0,30	0,30	0,7%
South Afr	0,000	0,000	0,06	0,18	0,4%
Mexico	0,000	0,000	0,05	0,05	0,12%
Europe	0,000	0,000	0,002	0,01	0,03%
Spain	0,000	0,000	0,000	0,01	0,02%
France	0,000	0,000	0,002	0,000	0,0%
Portugal	0,000	0,000	0,000	0,001	0,0%
Romania	0,000	0,000	0,000	0,002	0,0%
Ukraine	0,000	0,000	0,000	0,001	0,0%
Total	2,601	11,510	28,623	41,480	100,0%

GM crop area by trait: Of the 41.5 Mio hectares sown with transgenic crops in 1999, the distribution of traits in order of importance is as follows.

— herbicide tolerant (HT) GM crop with 69% of total,

— insect resistant (IR) GM with 21%,

— GM crops containing both genes (HT+IR) represented 7%

— and virus resistant (VR) GM crop (almost exclusively Chinese tobacco) nearly 3%.

Although this is the same order as in 1998, the area of crops containing both genes, the herbicide tolerant and insect resistant, has increased.

Of the 41.5 Mio hectares sown on a commercial basis in 1999, 53% were soybeans, 27% corn, 9% cotton, 8% rapeseed, 2% tobacco and 0.1% potatoes. Soybeans and corn account for

80% of GM areas world-wide. The area development of these most frequent GM crops is specified below.

GM soybeans: Commercialised GM soybeans were first sown in 1996 in 2 countries, the USA and Argentina and represented respectively 1.6 and 0.8% of their total soybean area.

Table 2: Development of GM soybean area

Mio ha	1996	1997	1998	1999	2000(e)	GM %('99)
USA	0,40	3,64	10,12	15,00	15,2	51%
Argentina	0,05	1,40	3,43	5,50	5,9	75%
Canada		0,001	0,04	0,10	0,1	10%
Brazil				1,18	1,3	10%
Romania				0,001	0,001	NR
Total	0,45	5,04	13,59	21,78	22,5	47%

In 1999, GM soybean area represented nearly one third of world total soybean area and nearly 47% of area of countries producing GM soybeans. Of the 22 Mio ha, 15 or two-third of total are in USA, 5.5 in Argentina, 1.2 in Brazil and less than 0.1 Mio ha in Canada and Romania. Almost all GM soybeans are herbicide tolerant (HT).

GM corn: First sowings of GM corn took place in 1996 exclusively in North America, 0.3 Mio ha in USA and 0.001 Mio ha in Canada and represented respectively 1% and 0.1% of their corn area.

Table 3: Development of GM corn area

Mio ha	1996	1997	1998	1999	2000(e)	GM %('99)
USA	0,30	2,27	8,66	10,30	8,8	36%
Argentina		0,07	0,09	0,31	0,7	11%
Canada	0,001	0,27	0,30	0,50	0,5	44%
South Afr			0,05	0,16	0,5	5%

France				0,002	0,000	0,0%
Spain				0,01	0,0005	0,2%
Portugal				0,001	0	0,4%
Total	0,30	2,61	9,11	11,28	10,5	28,0%

In 1999, GM corn sowings accounted for more than 11 Mio ha and 27% of total GM sowings. With this area, GM corn represents about 8% of world total corn area and 28% of area of countries producing GM corn. Most of the areas are located in USA, 0.3 Mio hectares in Argentina, 0.5 in Canada and a few thousands hectares in Spain, France and Portugal. Two thirds of corn area or nearly 8 Mio hectares are insect resistant (Bt-corn), about 2 Mio ha is herbicide tolerant corn and around another 2 Mio ha of corn contain both genes.

Soybeans and corn are well adapted to growing conditions in Northern and Southern America. Thus, they are widely grown in this part of the world, while culture in Europe is limited. The fact that corn and soybeans were the first species for which GM varieties were put onto the market is one basic factor explaining their development on the American continent. However, many other factors explain the rapid and uneven development of GM sowings throughout the world.

The marketing strategy developed by biotech firms has been focused on farmers, the first customers interested in agronomic traits of GM crops. They have shaped farmers profitability expectations. In the case of herbicide tolerant crops, the marketing strategy was based on the concept of "technological package". Many biotech firms are selling both the GM technology/seed and the associated crop protection product. This allows for "combined marketing", including adjusting prices of seeds and chemicals and using the same distribution channels. When selling their technology, biotech companies are charging a "technological fee". It results from the private origin of the new technology and has to be considered together with property and patenting rights.

Generally, the technological fee is first paid by seed firms, and is later transferred to farmers. GM seeds are sold in the framework of contracts which generally preclude seed-saving by farmers. The technological fee and the restriction on seed-saving imply increased seed costs- as such costs are to be paid each year- and a loss of autonomy for farmers. For Bt corn, "although analysis provides suggestive rather than conclusive evidence" they consider that "seed companies capture a significant, but by no means all of the net revenue advantage of Bt corn" and that "the likelihood of monopolistic pricing of the technology appears limited". For HT corn, they showed the sensitivity of profitability results to both the price of seeds and of herbicides, hence the sensitivity to the "combined pricing" strategy of the firms.

As far as HT soybeans are concerned, the American Soybean Association (ASA) has recently complained about significant differences in prices of Round Up Ready soybean seeds between the US and Argentinean markets. According to ASA, a bag of such seeds costs 12 US $ more in the US, and part of this difference is attributable to the 6 US $ technological fee, which is apparently not charged in Argentina. The combined pricing strategy and the observed variations in GM seed prices point to the existence of margin of manoeuvre of biotech firms for the price of technology. The market power of seed and agro-chemical suppliers deserves further assessment.

Biotechnology has generated increased concentration on the input side of the crop sector. This raises the question of increased dependency of farmers on a limited number of suppliers for crop production. Moreover, some biotech firms have already concluded agreements with grain processors, as is the case with the Monsanto/Cargill cluster. As a result, biotech appears as a driving force for vertical integration and for further consolidation throughout the agri-food sector. The downstream side of the food chain is also quite concentrated, either at the level of food processors (US) or at the retailing industry. The position of farmers in a rapidly changing agri-food sector is an issue of concern. The risk for them is to be "squeezed" between two oligopolistic industries.

Nevertheless, more and more contracts are governing the supply of crops by farmers, from the seed to the wholesale or processing stages. Biotech is very likely to be a driving force in such a process, for two reasons.

— GM seeds are sold and sown under contract. GM crops require adjustments in growing and management practices.
— If segregation or identity-preservation develop, crops, be they GM or not, will increasingly be grown and sold in the context of contracts. For this reason, some farmers are considering GM crops as "another liability".

The adoption of GM crops by farmers in the US, Canada and in Argentina has proceeded at an unprecedented rate compared to the uptake of conventional hybrids.

The analysis is based on the available economic literature, which mainly concerns Northern America. It is limited to the two main GM crops under cultivation Herbicide-Tolerant (HT) soybeans and Insect-Resistant (Bt) corn. Two Canadian studies on HT Canolahave also been taken into account.

FACTORS OF PROFITABILITY

Profitability is defined as the margin left over to farmers when costs have been deduced from receipts. The profitability of GM crops is judged against corresponding conventional crops. On the one hand, GM crops are expected to allow for saving in costs through different/reduced pest control and/or to achieve higher yields. On the other hand, GM seeds are more expensive than conventional ones. Under the assumption that the price of non-GM and GM crops is the same, the latter will become more profitable for farmers if the increased seed costs are offset by savings in pest control costs and/or by higher yields. Therefore, yields, seed and pest control costs are key factors for the profitability of GM crops.

GM seeds are sold at a higher price than conventional ones: While convergent figures could be found for HT soybeans, various figures are reported for HT Canola and Bt corn, depending on trade-mark varieties. The most frequently cited figure for Bt corn is a 22 /ha premium. The price wedge is mainly attributable to the "technological fee", but it also reflects the fact that markets for both types of seeds are separate. As far as weed and insect control is concerned, the situation is different for HT and insect-resistant (Bt Corn) crops. HT crops appear to allow for savings in herbicide costs.

However cost differences between biotech-based and conventional weed control programmes are not clear-cut and there are wide margins of fluctuations. While the total use of herbicides associated with HT crops has increased, the use and price of other herbicides have decreased. According to USDA, the net effect is a decrease in herbicide use. The herbicide effect of HT crops deserves further assessment, both on farm-level and globally, based on the experience of several years of cultivation.

Based on 1996-1998 data of the Agricultural Resources Management Data, the USDA has observed that adopters of Bt corn had obtained higher yields than nonadopters. However, this might partly be explained by performance differences between these two groups of farmers.

The yield of HT soybeans appears to be lower than for conventional varieties. A factor of explanation might be that HT genes have not been incorporated in top-yielding varieties. Comparing yields of GM and non-GM crops is not a straightforward exercise. Yields depend on a large number of factors, and the inserted trait of GM crops is only one factor amongst others. It is worth recalling that first generation genetic modifications address production conditions, they do not increase the intrinsic yield capacity of the plant. Not surprisingly yield performance of GM crops against their non-GM counterparts depends on growing conditions, in particular on the degree of infestation in insects or in weeds, hence on region of production.

Data about yields of GM crops are widely available, however, often specifications on factors which influence yields are missing, such as temperature, weed control applied etc.

The available studies do not provide conclusive evidence on the effective profitability of GM crops:

— HT soybeans: when comparing returns per ha or per labour unit, no significant difference appears between HT and conventional crops.

— The cost-effectiveness of Bt corn depends on growing conditions, in particular on the degree of infestation by corn borers. Results regarding profitability are contrasted, none can be considered as significant.

— There are no clear-cut results allowing for comparing the profitability of HT Canola with non-GM crops.

These rather contrasted and unclear results indicate that short term profitability is not the only driving force for adoption of GM crops by farmers. Other factors must have played a significant role in the rapid extension of GM sowings. In practice, the most immediate and tangible ground for satisfaction appears to be the combined effect of performance and convenience of GM crops, in particular for herbicide tolerant varieties. These crops allow for a greater flexibility in growing practices and in given cases, for reduced or more flexible labour requirements.

This convenience effect should translate into increased labour productivity and savings in crop-specific labour costs. However, this effect is not always assessed in profitability studies. One author concludes that HT soybeans provide the same returns on ha or on labour as conventional crops. But if they allow for reduced labour costs, the same return on less labour means increased profitability. This convenience effect has to be further assessed in particular, the valuation of the labour effect. For the time being, it does not translate into increased profitability, but rather in terms of attractiveness of GM crops for efficiency purposes.

The USDA has examined different factors affecting the adoption of HT soybeans and concluded that "larger operations and more educated operators are more likely to use the technology". It is very likely that the same applies to Bt Corn. The decision to plant Bt corn is a complex one, it implies assumptions as to the expected degree of infestation, adjustments in planting planning to foresee refuges. Such differences between adopters and non-adopters of biotechnology have to be taken into account when comparing yields and returns obtained on both types of farms.

The higher degree of education might echo the skills required for changes in growing and management practices. The farm size of adopters might be a factor explaining, amongst others, the dramatic increase in areas sown to GM crops. The adoption of biotechnology is not size-neutral. The reviewed studies only compare farm-level and short-term profitability. Profitability of GM crops should be analysed over a longer timeframe. First, there are important yearly fluctuations in yields and prices, and it is difficult to isolate the possible effect of biotechnology. Results are very sensitive to the price of seeds and agro-chemical products on the one hand and to commodity prices on the other hand. In most profitability studies, prices for GM and conventional crops are assumed to be equivalent. Developments on the supply and on the demand side of the food chain have to be considered together, and this is another reason for assessing profitability over several years.

In the case of HT crops, gains in efficiency should translate into improved labour productivity. In the case of Bt corn, yield gains mean enhanced productivity of land. Both types of effects imply a shift in farmers supply functions. Under given prices, enhanced farm productivity leads to an increase in supply. If the demand function remains unchanged, prices drop. In the long run, enhanced productivity will have an impact on farm restructuring, alongside with many other factors playing a role in this process. While more and more producers are adopting biotech crops, thus contributing to the increase in supply, on the demand side, concerns about GM food are emerging. As a result, segregation between GM and non-GM crops is developing, which implies differentiation in costs and prices.

Agricultural Policy Effect

In the US as well as in the EU, GM and non-GM crops are not treated differently under the various support schemes, both are eligible. In the US, crops for which GM varieties have rapidly developed are all eligible for support under the flexibility payments, the marketing loan system, as well as for crop insurance. Soybeans became eligible for flexibility payments and under the marketing loan system in 1996, which is the year of first commercial sowings of GM varieties. Several analysts consider that existing support systems have favoured the development of soybeans sowings. In particular, the loan rate applied to soybeans makes this crop attractive compared to wheat and corn. The area under soybeans is expected to reach a record level in 2000, while prices are low.

Favourable support conditions for soybeans could have played a role in the rapid uptake of GM technology for this crop. In addition, in a low market price context, the expectation on cost savings is a further driving force for the adoption of the technology. Eligibility of GM crops under various support schemes limits the price risk of the productivity-enhancing technology. It accounts as another reason for the farmers to focus their planting decision on expected farm-level performance, on cost-efficiency of inputs.

SEGREGATION AND IDENTITY PRESERVATION

The development of efficiency enhancing GM crops dominates the agricultural applications in most countries where GM crops are grown. The EU debate on GMOs, on the other hand, is dominated by demand factors, such as food safety concerns. Consumer reaction to GM food has given rise to uncertainty about market developments, both as regards to the short term prospects for GM products and to the future competitiveness of conventional non-GM production.

Trade of agricultural products today is based on the commodity system, which works on the basis that crops from different farms are sufficiently alike to be traded at a common price and to a common grading specification. On its journey to a milling plant, the crop can be sampled and blended several times and there is no traceability back to the producer. Bulking up the produce of many producers means that transport and handling costs can be reduced. Furthermore, bulk transport enables a continuous flow for processing, since taking a processing plant down and firing it up again can be time consuming and costly.

Segregation refers to a system of crop or raw material management which allows one batch or crop to be separated from another. It implies that specific crops and products are kept apart, but does not necessarily require traceability along the production chain. Identity Preservation (IP) is a system of crop management and trade which allows the source and/or nature of materials to be identified. The objective of IP is to ensure that a particular crop is monitored throughout the food chain and thus to guarantee certain traits or qualities which might command a premium. IP requires a set of actions to allow traceability and is usually communicated to the consumer by a label. Currently IP is used to identify crop varieties which provide additional features concerning the content or composition of products. There are several reasons to consider IP systems in the GMO context:

— consumer request for traceability in order to control health and environmental effects,

— international agreements, in particular the Biosafety Protocol,
— mandatory GMO-labelling requirements in certain countries,
— differences in approval status of GMOs in different countries,
— consumer demand for certified non-GMO or GMO-free products, and
— the development of GMOs, with specific traits addressing the consumer and the processing industry.

Segregating and Identity Preservation are attempts to create and establish separate markets for differentiated products or to set up a "new" market for a "new" specific crop. The economic effects of segregated markets for GM and GMO-free crops correspond to a dis-aggregation of the supply and demand curve. Assuming that the aggregated supply for a certain crop would be subdivided equally among GM and non-GM markets and demand would follow the same pattern for both sub-markets, the price should be the same on the GM and on the non-GM market.

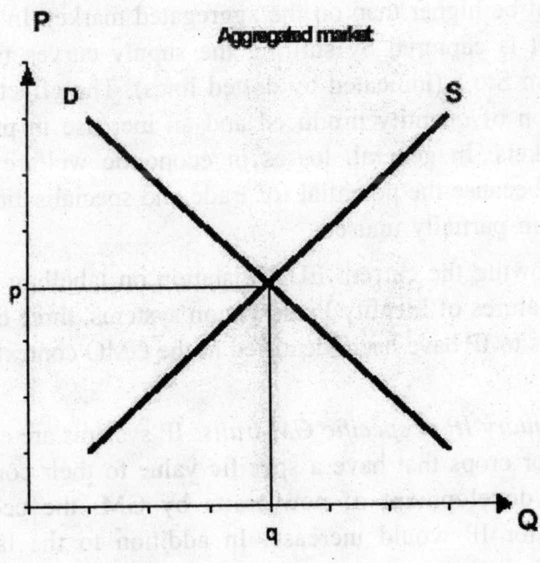

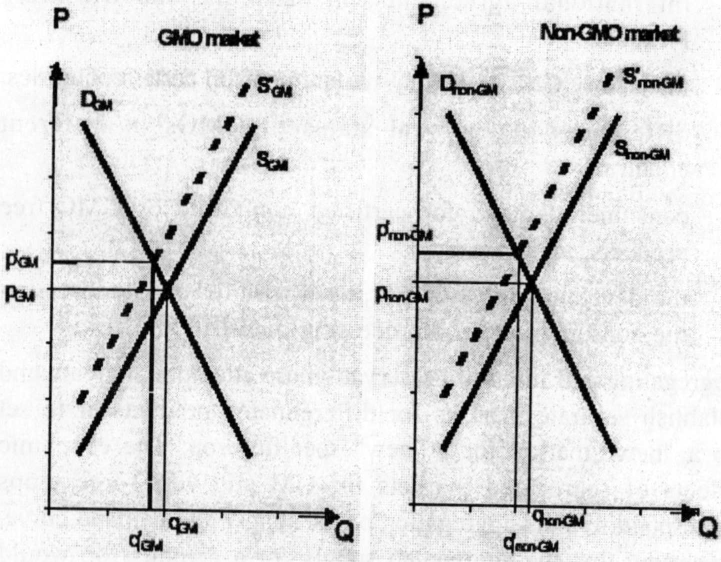

Figure 1: Segregation of GM and non-GM markets

However, due to lower quantities produced and traded, potential economies of scale may not be used and production cost per unit might be higher than on the aggregated market. In figure 1 this effect is captured by shifting the supply curves from Sto S'and from Sto S'(indicated by dotted lines). The effect will be a reduction of quantity produced and an increase in prices on both markets. In general, losses in economic welfare can be expected because the potential for trade and specialisation gains will remain partially unused.

Following the current EU legislation on labelling and the general features of Identity Preservation systems, three different approaches to IP have been identified in the GMO context (figure 2).

Voluntary IP of specific GM traits: IP systems are common practice for crops that have a specific value to their consumer. With the development of new traits by GM, the economic incentive for IP would increase. In addition to the labelling requirements under the novel food regulation, there would be a clear incentive on the supply side to introduce IP and thus to preserve the additional value or quality of such a GM crop through the processing chain. IP would distinguish a product for which consumers are expected to pay more than for a conventional product.

Voluntary IP of GMO-free products: The second approach for IP is to preserve and label GM-free products in order to enhance consumer choice. Current EU legislation already requires compulsory labelling for food containing GMOs. Thus, the introduction of labelled GMO-free food would in theory enable the choice between three categories of foodstuffs: novel GM food, conventional non-GM food and GMO-free products. However, it can be expected that the share of conventional food will diminish over time, since the pay off for GMO-free products can be expected to be higher than for conventional non-GM products. If producers decide to make an effort to segregate, the additional costs to comply with GMO-free standards might be low compared to the additional premia achieved on the market.

On the other hand, if at least part of the consumers accept labelled GMO food, some conventional raw material would enter into GM-labelled final products if voluntary IP of non-GM products is not rewarded by a sufficiently high premium.

Compulsory IP for GM products (GM traceability): Trading GM crops through commodity systems prevents material being traced within the transportation and processing chain. Thus any commodity sample originating from a region or country where GM and conventional crops are grown in parallel might contain GM crops. Traceability, i.e. a compulsory IP system, has been introduced as a strategy to re-establish consumer confidence in the EU beef sector following the BSE crisis. Traceability could also be a strategy to monitor the environmental and health effects of GMOs and to enable choice to those consumers who want to avoid GMO consumption.

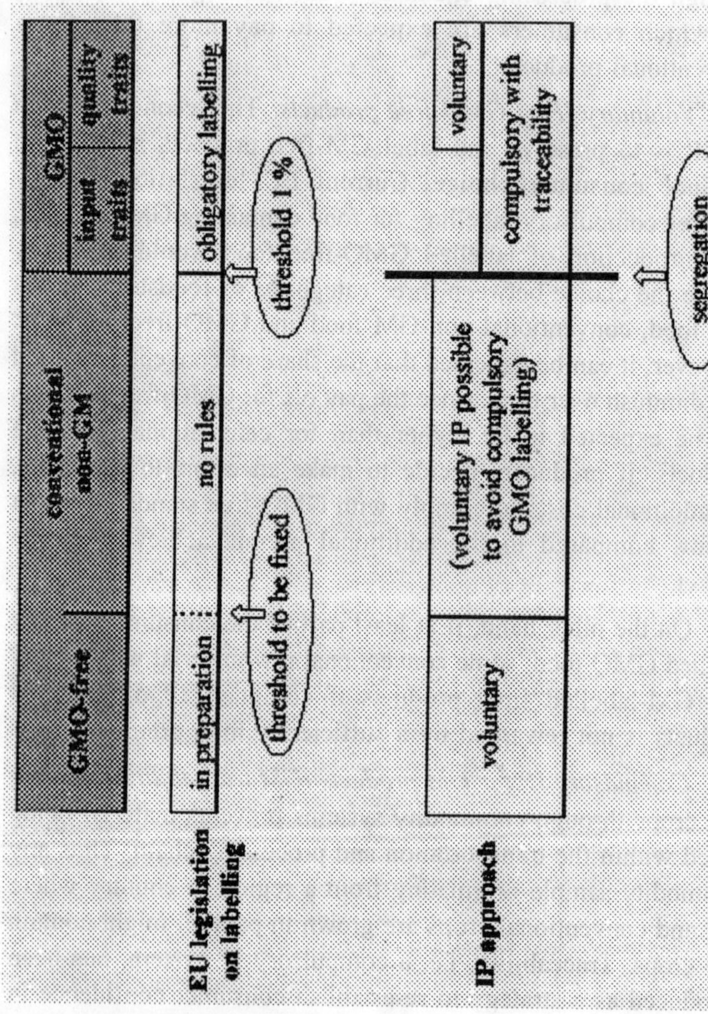

Figure 2: Labelling and Identity Preservation

Methods of GMO Testing

A Genetically Modified Organism can be distinguished from a non-GMO by the fact that it contains either unique novel deoxyribonucleic acid (DNA) sequences and/or unique novel proteins not present in its conventional counterpart. Two methods are actually applied:

Polymerase Chain Reaction (PCR): The polymerase chain reaction is based on the detection of DNA fragments that are inserted in the plant genome. This method allows amplification in a few hours of specific DNA fragments to a degree that they can be analysed qualitatively and quantitatively by common laboratory techniques, e.g. electrophoresis. However, it requires specialised equipment and training. PCR testing is applicable and extremely sensitive in the case of unprocessed food where the DNA is still intact. This is not the case for processed food where it is more difficult to isolate high quality DNA and where GM material from more than one GM species can be present. PCR requires little reagent development time, but it can still take 1 to 3 days to receive results from a testing laboratory. The test is estimated to be about 99.9% accurate.

ELISA: This method is able to detect and to quantify the amount of a certain protein which is of interest in a sample that may contain numerous other dissimilar proteins. ELISA uses antibodies to bind specific proteins, for instance those newly synthesised following a genetic alteration. A colorimetric or fluorometric reaction can visualise and measure when the antigen and specific antibody bind together. One restriction for using the ELISA test is the denaturation of proteins in some food processes. This method also requires high investments to develop the assay and to generate antibodies and protein standards. However, once reagents are developed, the cost per sample is low. The test is reported to be 95% reliable.

GMO Costs of Identity Preservation

Additional costs of IP arise with the additional work involved in growing, handling, storage, transport, processing, cleaning, and administration.

Seed production: Already under conventional systems basic and certified seed is normally distributed separately bagged and labelled. The crucial variable to determine additional costs in seed production will be the tolerance level for contamination with other (GM) varieties. Representatives of the seed industry have confirmed that they could provide seeds at any desired tolerance level. However, costs would rise following rather an exponential than a linear function with a tolerance level approaching zero percent.

Farm production and on-farm handling: The farmer will be able to control the likelihood of volunteer plants, mechanical commingling and the distance to avoid cross pollination. Physical distance between the pollen donors and the crop is the most important factor to avoid cross pollination among specific varieties. Cross-pollination raises a number of legal and economic issues concerning the coexistence of different production systems, i.e. GM, conventional and organic farming. For instance, the standards for organic farming provided by the UK Soil Association require minimum distances from GM crop plantings. The crucial factors to determine IP costs at the farm level will be the tolerance level to be achieved, the physical ability of cross pollination and rules and legislation concerning neighbouring farms. However, most of the additional costs at the farm level would be avoided, if the full production could be switched to a single type of IP.

Testing: For GM crops providing quality traits testing will refer to these specific modifications. GMO traceability would extend the need for testing to all genetic modifications, including agronomic traits. For GMO-free products, the testing would not be limited to determine the presence or absence of GMOs, but would also have to confirm that the tolerance levels have been respected. In order to compare the different cost elements, testing costs have been calculated per tonne, although testing is not only applied to raw products but also to processed foodstuffs. The additional cost per tonne of soya or corn for testing the presence of a specific biotech trait by the ELISA technique has been estimated at 0.4

However, since current ELISA testing methods require a separate test for detection of each unique trait, several tests may be necessary. At subterminals and export elevators, PCR testing is more common than ELISA because it is more sensitive and can be used to detect presence of several genetic modifications by one set of tests. Furthermore, it becomes more efficient with larger volumes of grain to be tested. Overall cost for an IP testing system have been estimated to range from 1 /t for a simple checking to as much as 20 /t for the most disciplined systems of overlapping documentation, field inspections, product sampling and laboratory testing by third parties.

An alternative to expensive tests could be the introduction of additional genes that provide visual markers to facilitate identification. However, IP documentation is likely to reduce the need for testing compared with on the spot testing of commodities for GMO contamination or specific traits.

Transportation and storage: In general, increased IP trade would reduce the value of the traditional commodity infrastructure. Additional costs will occur with the need to find separate storage at local elevators with cleaning and with possible restrictions in the delivery schedule. If transportation and storage facilities in silos, trains, trucks or ships cannot be fully used by IP crops, further costs might occur per unit. According to the literature available, the additional transport cost range from 1 to 9 /t for the different products and IP approaches. These costs represent about 0.5 - 5% of the farmgate price. The key factors will be the amount of crop traded under the different IP systems and the tolerance level for contamination.

Processing industry: Normally, processing plants for soybeans and corn are run continuously except for annual cleaning or repair breaks. Stopping production and cleaning the facilities would cause additional cost. The specific cost of IP processing further depend on the number of secondary products produced from the raw material. If only one of the output products is required to be IP, e.g. the soya oil, it will bear the whole cost of IP. If there is a market for all the products of IP however, then the costs of IP will be spread across all end products. If there is sufficient IP supplies of a crop, it may be possible to dedicate a plant to processing such supplies, in which case there would be no additional costs involved from separate processing and storage.

Total cost of IP systems: Summarising the different costs along the production chain allows estimating the total costs of IP. According to the literature available, they range from 5 to 25 /t depending on the different grains and the IP systems. Thus, IP would increase the grain price by 6 - 17% compared to the farmgate price. Since such a range corresponds to the experience with well established IP systems for value added market segments, it can be taken as a reliable estimation of IP costs.

The following main factors have been identified to determine IP costs:

— *Tolerance level:* The more stringent the purity requirements, the more expensive will be the IP system. The tolerance level appears to be the most important cost determinant for all three IP approaches discussed in this report. Fixing a threshold will particularly concern the cost of seed production, the costs for testing, storage and transportation and the decision to switch a whole farm and a whole processing plant to specific (IP) production. Choosing a severe level of tolerance may increase the cost to such a high level that they would override the possible benefits of IP production. An extremely low tolerance level for GMO-free products could thus be a strong disincentive to establish GMO-free production and would reduce the GMO-free market to niche production for high income households.

— *Agronomic traits:* The genetic disposition for cross pollination and for volunteers will determine in particular the costs on the farm.

— *Market volume:* Economies of scale can be expected for any IP system. The more crops are traded under such a system, the higher will be the potential to reduce costs. Furthermore, if an entire stream can be devoted to an IP system, additional costs should be quite low.

— *Seasonality:* A strong seasonality of market supply could increase the storage costs of an IP system, in particular if the IP crop is grown only in a particular region or country.

— *Derived products:* IP costs per unit depend on the share of all processing products which can be marketed as IP. If only one of a whole range of the output products is to be identity preserved, it will bear the whole costs of IP.

Nevertheless, the magnitude of the additional costs is not fixed. It depends on the particular circumstances. Additional costs for segregation and IP systems have been sown to occur in the different stages of the production process. However, these costs can be shifted between the different stages along the chain. Analysing their allocation is important to understand the economic effects of IP. Four factors, which determine the sharing out of costs have been described as:

— *Price responsiveness (own-price elasticity):* Depending on the responsiveness of demand and supply to price at each of the stages additional costs can be shifted - at least partially - to the previous or to the following stage of the production chain. Generally the less price-responsive demand is at a certain stage, the more of the additional costs will be absorbed by the consumer at this stage. Equally, the less price-elastic is supply, the more of the additional costs have to be absorbed by the producer.

— *Availability of substitutes:* The more substitutes are available, the more responsive would be the price. Thus for products, which can easily be substituted, additional costs will hardly be shifted to the processor or the final consumer. In this case, it will be the farmer who has to bear most of the additional costs of IP. On the other hand, if a product is difficult to substitute, it will be the consumer who has to bear the IP costs.

— *Market structure:* Price-responsiveness can be affected by the competitive structure of the industry. The more concentrated the structure, the more likely that any additional costs are passed over to the previous or the next stage of the chain. In the food sector, the market power is in general stronger at the food processing and retailing levels compared to the farmer and consumer level. Thus IP costs are very likely either to be passed back to the farmer through lower prices for his products or to be passed forward to the consumer in the form of higher food prices.

— *Agricultural price policy:* Agricultural policy measures, in particular those established to control agricultural prices may have an adverse impact on the transmission of additional costs to the consumer. On the other hand, price policy may also reduce the transmission of benefits of cost reductions by new technologies and thus reduce the economic incentives to apply these innovations.

These factors apply to all three IP approaches which have been identified in the context of GMOs.

Voluntary IP of specific GM traits: If GM crops have a specific value to the consumer, these crops have to be handled separately, in order to preserve their value through the chain. Price elasticity of supply can be expected to be high. On the demand side, the new trait will create a situation in which the scope for substitution is limited and thus demand gets fairly price inelastic. The effect will be that most of the additional cost can be passed on to the consumer. The market will be a niche market - at least in the beginning - for each of the new traits introduced by genetic modifications. Thus it is very likely that the consumer will be charged a premium which covers not only the intrinsic additional value of the new product, but also the costs to handle them separately through the food chain.

Voluntary IP of GMO-free products: If GMO-free products have a specific value to consumers, they are willing to pay a premium for these products, which are handled separately or identity-preserved. With a voluntary IP system for GMO-free products, additional costs will be borne by the producers, processors and consumers of these GMO-free products. The scope for passing over the costs of IP for a GMO-free product will depend upon how strong the demand for GMO-free products will be. The stronger the demand, the less responsive will it be to price change. This would increase the scope for suppliers to pass over the costs of IP in the form of higher prices. Thus it will be more likely that the consumer bears the costs than the farmer of GMO-free crops.

For the short-term development, however, some impact on the market for GM crops cannot be excluded. In a short-term analysis supply of GM and GM-free products is assumed to be fixed. Consumers without specific preference for non-GMO products will not care whether they consume GMO or GMO-free products. However, GMO-free demand will not accept GMO supply. So there will be one-way situation for substitution and the magnitude of demand for IP products relative to the demand for commodities will be the crucial factor to determine the distribution of the additional costs as well as of the price of GM and GMO-free crops.

To analyse the short term market effects, two scenarios can be distinguished:Price information for soybeans indicate that US producers have received a premium of 5 - 9 /t for non-GM soybeans in the last years. This amount corresponds to about 4% of the farmgate price. More recent sources signal a lower premium level of 3 - 7.5 /t. In contrast, GM soybeans are being discounted by up to 10% of the farmgate price in many parts of the USA. However, according to US grain handlers, the premium paid for food quality soya was much higher than the non-GM premium. The premium for organic soybeans was estimated at 245 /t, a premium of almost 150% of the commodity price.

Thus, farmers who are thinking about entering into non-GM production might consider as well to switch to high quality varieties or to organic farming in order to realise the higher market price. While quality trait premia for corn range between 4 and 6 /t, non-GM premia appear to be slightly lower. They range between 1.8 and 5.6 /t. IP premia range between 2.5 and 9% of the farmgate price for corn. However, when these price differences per tonne are translated into price differences per hectare the farmer will have to take account of yield differences. Yields of quality trait varieties are often lower than average, while several studies have found evidence on yield gains for Bt corn compared to conventional varieties.

Compulsory IP for GM products: Since most of the quality traits introduced by genetic engineering can be expected to rely on voluntary IP to preserve the additional value, GMO traceability would mainly affect crops with modification of agronomic traits. Agronomic traits address the producer and the crops are marketed similar to conventional crops. Thus any consumer without particular preference for GMO-free food should be indifferent when comparing GM and GMO-free products. A high degree of substitutability can be supposed, because the consumer could easily switch completely to the conventional product if additional cost for IP would increase the price of a product.

This would mean that IP costs would be passed back to primary producers and processors of GM crops. The producers of conventional crops would not be affected and the additional IP costs at the farm level would reduce the profitability of GM crops. The relative position of GM and conventional crops could be altered, if the agronomic trait is sufficiently advantageous at the farm level. As soon as the GM crop accounts for a significant proportion of all traded crops, it becomes the norm and will set the baseline for the commodity price of this crop. This would reduce the competitiveness of conventional crops and increase the incentive to adapt the production programme.

Markets for Soybeans and Corn

Soybeans: The EU is the world's leading importer of soybeans and soymeals. Domestic production of soybeans is covering only a small percentage of EU consumption. Most soya-bean/meal production and imports are used for animal feed, but a small share is used for food. The EU main - and nearly exclusive - trading partners for soya beans and meal imports are Brazil, Argentina and the US. The European market is of particular importance for Brazil and Argentina. 40 to 50% of their soya production is sold to the EU. The USA as the world's leading soybeans exporter, are sending 10 to 15% of their production towards the EU, which, is equal to around 30% of USA soya exports. Thus, for soya bean and meal trade, there is a mutual dependency between the three main exporters and the EU as the main importer. Given this mutual dependency, and taking into account that:

- more than 50% of the US soybean area and almost three quarter of the Argentinean soybean area are under GM crops,
- segregation of GM and non-GM crops is still limited in the US and there is no evidence on segregation in Argentina, it is very likely that animal feedstuff in the EU consisting of or containing soya imported from these countries contain GMOs. Soymeals represent an important source of proteins for poultry and pigs. Therefore it must be assumed that currently most chicken and pigs fed in the EU have already eaten some GMOs.

Corn: In corn production, the EU has reached a degree of self-sufficiency which is around 100%. Imports contribute 4 - 8% to total availability on the internal market. Feed use absorbs about 75 - 80% of the EU market volume, industrial use accounts for 4.2 Mio tonnes each year (11-12%), and human consumption for 2.6 Mio tonnes (7%). However, imports of corn by-products, in particular corn gluten feed, surmount the imports of corn grains. In 1999, around 4.7 Mio tonnes of corn gluten feed was imported by the EU. For corn the USA is the worlds leading producer and exporter, although only 20% of the US corn production is exported. The main part is sold on the domestic market for feed (60%) or non-food uses. EU imports of US corn have decreased dramatically. The share of US in EU corn imports dropped from 86% in 1995 to 12% in 1999. Meanwhile Argentina has become the major supplier for EU imports.

While accounting for the main producer of GM corn and soybeans, the US are the leading exporter for these commodities. Argentina is the second biggest producer of GM soybeans and the third exporter. The main importing countries for these commodities, the EU and some South-East Asia countries, have taken a restrictive stance on GM food. In particular, labelling of the GM nature of food ingredients is compulsory in the EU. Japan intends to implement mandatory labelling by the second half of 2000. Not surprisingly, this situation has become a trade issue. However, it is difficult to isolate the possible effect of biotechnology on developments in trade, as many other factors play a role, like changes in competitiveness, transportation costs and the transaction costs of giving up of long-established trade links.

US soybean exports declined from 26 to 20 Mio tonnes between 1997 and 1998, while world soybean trade held fairly steady. EU soya imports from the US have been partially replaced by imports from Argentina. The USDA has concluded that "traditional competitive forces appear to be the main driving factors behind the changes in observed bilateral trade patterns". As the share of GM soybeans is much higher in Argentina than in the US, this shift in trading pattern cannot be attributed to reluctance to import GM soybeans. The drop is even sharper for corn than for soybeans. US corn exports fell from 60 Mio tonnes in 1995 to 41 Mio in 1998. Most of the drop occurred on South-East Asia markets and is explained by the situation of China, which became again a net exporter of corn.

On the EU market for corn, the share of US has steadily fallen while the share of other partners, in particular Argentina and Hungary, has significantly increased. The USDA considers that the loss of shares on the EU market results from issues related to biotechnology, in particular the differences in regulatory approaches. While 11 types of GM corn have been approved in the US, only 4 have been cleared at EU level, and some Member States have decided to suspend authorisations for growing. Non-authorised GM crops cannot be placed on the EU market. In the absence of tolerance thresholds, if traces of such crops are found in a given consignment, it cannot be cleared for importing into the EU. However, the type of GM soybeans which is mostly grown in the US is authorised in the EU for imports and processing. Only a small part of US areas have been sown to non-EU approved corn varieties and the EU only accounts for 1% of US corn exports.

Trade issues have been addressed in the Biosafety Protocol, which aims at ensuring an adequate level of protection for transfer, handling and use of LMOs which might have an adverse effect on biodiversity. Reference is made to the precautionary principle in this respect. It is hoped that procedures foreseen under this Protocol, in particular information sharing and accompanying documentation, will help improving the predictability of transboundary movements of GMOs.

Chapter 9

Biotechnology in the Developing World

The role of agriculture in food security extends far beyond growth in crop yields and total production. Agriculture promotes food security primarily when it contributes to incomes and productive employment. Enhancing food security in the poorest regions requires investments in scientific research and training and the transfer of knowledge and technology to ensure wise management of resources and sustained capacity for growth. Moreover, food security dictates a focus on poor people's crops: subsistence and marketed crops grown in marginal areas where the poorest segments of the rural population are concentrated. Few tabulations of the importance of orphan crops exist.

Wheat, rice, maize and soybeans each occupy more than 70 million ha globally per year. Collectively, they cover 580 million ha and generate on the order of US$ 300 billion in gross value annually. Approximately two thirds of their combined area is within developing countries. Remarkably, these four crops also supply an average of about 1360 kcal of energy and 33 g of protein daily to individuals in poor counties.

Orphan crops also play a large role. For example, millets are key crops for the poor in both Sub-Saharan Africa and South Asia. In addition to grouping sets of species like the millets.

Twenty-seven orphan crops within developing countries occupy areas of between 0.5 and 38 million ha. None is on the scale of wheat, rice, or maize, yet they total some 250 million ha. In addition, there are 70 million additional hectares planted to fruits and vegetables.

Orphan crops take on even more significance within regions. The data for Sub-Saharan Africa illustrate this point. Sorghum and millets are more important than rice and wheat, both in area and in contributions to the diet. Similarly, roots and tubers play a dominant role, providing more than 400 kcal of energy per person per day.Given that 38% of Sub-Saharan Africa's total population is undernourished and that the number of undernourished children in that region is forecast to increase by 39% by 2020, the issue takes on added urgency. Important questions remain regarding the types of investments that are needed for individual crops or crop groupings. Minimum investments in biotechnology approaches for individual orphan crops are likely to be at least as large as those for major crops for a particular trait. As a result, the economic incentives to invest initially in biotechnology for major crops seem clear.

Once such investments have been made, however, incremental investments that extend the science to orphan crops could be extremely beneficial to some of the world's poorest populations, an opportunity often ignored.

Molecular Tools and Germplasm Improvement

Molecular approaches have been widely used to describe and manipulate crop genomes and to better understand the genes and biochemical pathways that govern traits such as yield, disease and pest resistance, abiotic stress tolerance, plant architecture, and quality. Many national agricultural research programs (NARs) in developing countries have also made some investment in molecular technologies as applied to the major crops, such as rice and maize, but the extent of that investment varies sharply across countries. Nations with strong NARs are generally those with strong economies, such as China, India, and Brazil. Even in those countries, however, biotechnology investments are rarely applied to orphan crops. More than a decade of investment has developed powerful technologies for major crops, making genetic information for crop improvement more accessible. With this technology, DNA-based approaches have been applied in two broad areas of molecular breeding.

The first, marker-assisted breeding, is aimed at enhancing the power of conventional genetic analysis and manipulation. The second is transgenics. The first step of molecular breeding uses molecular markers as tools to detect the extent and structure of genetic variation, providing insights into the diversity of crop varieties and potential contributions offered by their wild relatives. Diversity seen at this level can be used to inform the selection of parents with a high likelihood of producing novel progeny in a crossing program. Molecular tools are also used in the analysis of inheritance of key crop traits, including those that are subject to complex inheritance due to the involvement of numerous genes. Once the genetic basis of a trait is well understood, molecular tools can be used to select for specific desired genes or combinations of genes.

The second avenue involves the transfer of genes from one genotype to another. This approach, often referred to as "genetic engineering", may utilize man-made genes, or natural alleles. Thus, truly novel traits can be added to a crop. The two broad areas of molecular breeding converge when genetic engineering is applied to enhance the efficiency with which native genes are moved within a gene pool.

Genome Analysis and Molecular Breeding Approaches

All plant genomes have a great deal in common in their gene content, biochemical pathways, and chromosome organization. Different plant taxa have different versions of the same genes at a given position or locus in a genome, but the order of loci is conserved to varying degrees across even distantly related crops, a phenomenon known as synteny. Because plant genomes are somewhat similar, investment in "model" species, chosen because they are easily studied, has paid off. Models have been selected for each of the most important plant groups, and research has been accelerated through large-scale projects.

For example, the flowering plants can be divided into two main groups: monocots and dicots. The former includes the grasses, which in turn include the major cereal crops. The latter include the legumes, many roots and tubers, and vegetable crops. The genome of the model dicot species Arabidopsis thaliana has been fully sequenced, and a great deal of basic research has been done on this species. More recently, the draft genomic sequence of the rice (Oryza sativa) genome was made available, which is useful both for rice and its monocot relatives. Ninety-eight percent of the proteins found in maize, wheat and barley are also found in ric. Although the progenitors of rice and Arabidopsis diverged 150 to 200 million years ago, more than 80% of the genes that have been documented in Arabidopsis have also been found to have related genes in ric.

The genes involved in many biochemical pathways and processes are very similar across the plant kingdom. Functions such as gene regulation, general metabolism, nutrient acquisition, disease resistance, general defense, flowering time and flower development are largely conserved across taxa. Comparative mapping studies have revealed that the genomes of plant species within families are syntenous-that gene order is conserved for chromosomal segments. Given the similarities among crop genomes, particularly among plant species within a family, it seems possible that research on major crops or model species would benefit a substantial number of related crop species in the same families. The potential spillover benefits are likely to differ, however, across taxa and region depending on the particular biology of the species, the constraints faced by farmers, and seed systems.

Genetic Diversity

The use of molecular techniques for analysis of genetic diversity and the structure of germplasm, the first conceptual step in marker-assisted breeding, has been fruitful for many species. This type of research can lead to better conservation of crop genetic diversity, more efficient selection of accessions for phenotypic characterization, and the identification of useful variants for conventional plant breeding. Some molecular techniques, e.g., amplified fragment length polymorphism, or AFLP, can be applied to any species, without the need for crop-specific DNA sequence information. Such analysis can be conducted in any laboratory for which minimum levels cf basic training, supplies and equipment are available.

Even with the most seemingly straightforward techniques, however, it is difficult to achieve smooth and productive operations under the conditions encountered by many scientists in developing countries. Administrators often err in thinking that investment in facilities alone is sufficient for scientific success, and they neglect the substantial costs incurred for training, consumables, and the diversion of key personnel from field-based efforts. Even when researchers are able to produce goodquality molecular data, they may lack access to expertise in some of the analytical procedures needed to interpret the results.

This area of research provides a clear example of one in which spillovers from investments in facilities, human capital, and supplies for major crops could be quite large for orphan crops. Once researchers are well trained in the techniques and have the necessary equipment, they could apply their skills to a wider range of crops, or they could help to train others to analyze genetic diversity of orphan crops.

Marker Assisted Selection (MAS)

Marker assisted selection (MAS) is the identification of DNA sequences located near genes that can be tracked to breed for traits that are difficult to observe. MAS can be relatively straightforward for genes conditioning large phenotypic effects, such as strong effects on disease resistance, and its use for germplasm improvement is beginning to prove successful for some types of field applications for major crops.Two new rice varieties with bacterial blight resistance, derived from polymerase chain reaction, were introduced commercially in Indonesia in 2002. Rice lines carrying multiple resistance genes are being produced by several national programs following technical investments at the International Rice Research Institute (IRRI).

In "single large-scale MAS", selected genes are fixed by marker-assisted selection, and the constitution of the rest of the genome is determined by conventional field-based breeding. This approach has been used to develop virus resistance in maize, rice, and beans. In recent cost analyses of marker-assisted versus conventional selection, the cost of MAS for a recessive mutant gene associated with quality protein maize was found to compare favorably with the (difficult) conventional screen.

As basic research on key plant traits advances and as marker technologies continue to become increasingly simple and powerful, more applications of MAS are anticipated with the potential benefit of faster varietal release. Positive costbenefit results of MAS, and there are many similar unpublished examples from the private sector. In the published literature the cost-benefit results of MAS are mixed, and many publications only report work aimed at eventual application of MAS as opposed to field-level use of MAS for practical crop improvement.

Due to the relatively high cost of MAS compared to conventional field screening, MAS is likely to be practical for orphan crops in the poorest countries only if partnerships with well-funded research groups allow free use of existing technologies. It is important, therefore, to assess various traits and trait categories for which there has been some proof-of-concept in non-orphans for MAS before considering what would be required to move the benefits to a broader set of plant species.

In viewing cases in which genetic research has allowed for progress in MAS, it is important to distinguish between two types of traits: those conditioned by single genes with large effects showing simple inheritance or qualitative traits, and those showing complex inheritance or quantitative traits. Important plant traits that frequently show simple inheritance include some resistances to diseases and insects and some important characters that relate to plant growth and development. It is relatively easy to locate these genes with precision, and in many cases such genes have been cloned. It is often straightforward to select for these genes in a conventional breeding program, in which case MAS does not necessarily offer substantial benefits.

There are cases, however, in which markers are particularly beneficial; for example, when a disease resistance gene must be identified but the pathogen is not present at the breeding site, or when a gene is to be selected at the seedling stage for a trait that is only expressed later in development. Most important traits are governed by multiple genes, each having relatively small effects. These "quantitative traits" have been difficult to understand and to manipulate in conventional crop breeding programs. The term QTL, quantitative trait locus or loci, refers to the chromosomal regions of genes that control quantitative traits. While so-called anonymous QTL-associated markers are sometimes used for marker assisted selection in crop improvement, the utility of such markers can be limited by a high degree of imprecision in mapping desirable loci of small effects.

The identification of "candidate genes" makes it possible to localize desirable variants much more precisely. Candidate genes are genes known or suspected to be involved in conditioning the phenotype of interest, such as disease resistance. The inference that a particular gene contributes to a given trait may be strengthened if it maps to a chromosomal region that has been associated with the trait in QTL mapping experiments. Credible candidate genes co-localizing with QTLs have been identified for several traits, including quantitative disease resistance in rice, wheat, bean, and potato.

A number of research approaches have converged to allow some genes underlying QTLs to be cloned, and to set the stage for future QTL cloning in others. This process permits both the identification of potentially useful variants of agronomically important genes and the precise selection of those alleles found to be most useful. One of the approaches that puts QTL cloning within reach is the increasing availability of express sequence tags (ESTs), short segments of sequenced gene transcripts that provide information on genetic expression, function, and heritability. Tens or hundreds of thousands of ESTs are available for major crop species, but the number of sequences for orphan crops is quite meager.

Researchers can now use the combination of QTL mapping data and mapped EST data as a powerful tool for determining the genes that underlie quantitative traits. Sequence data on expressed genes and on plant and crop genomes are rapidly accumulating and present powerful tools for plant science, provided that scientists are able to access and exploit the data efficiently. The sequence data sets are, in themselves, imposing and potentially cumbersome. But with the help of increasingly powerful and friendly databases, biologists and breeders can now gain access to genetic information that allows them to identify and exploit natural variation for a wide variety of crops in ways that were previously not possible.

Regulatory Genes

The genetic complexity underlying quantitative traits often makes it difficult to seek traitmarker correlations that can be used for indirect selection. For instance, a great number of genes are involved in plant defense against insects and pathogens. This complexity may seem defeating, but it may be possible to improve complex traits by manipulating the genes that sit at the top of the regulatory hierarchy and control large numbers of other genes. In fact, there is growing evidence that many QTLs are explained by transcription factors-genes that control the expression of other genes. The identification and utilization of the "best" alleles of these genes is a promising avenue of research, both via MAS and via transgenic approaches.

Transformation experiments enhancing the stress-responsive expression of a transcription factor that controls drought-related genes led to increased tolerance to drought, salt stress, and freezing in Arabidopsis. Similarly, overexpression of a regulatory factor involved in plant defense in transgenic Arabidopsis led to enhanced resistance to diverse pathogens. Some of the genes involved in regulation of the defense response are also implicated in responses to other stresses. The identification of these transcription factors seems likely to play a useful role in genetic improvement for a wide range of food crops.

TRANSGENIC DELIVERY OF CROP PROTECTION TRAITS

The use of direct gene transfer, the second tool for "molecular breeding" of crop traits, employs recombinant DNA technologies to insert one or more genes into the crop plant's genome.While most transgenic research and applications have focused on a few major crops to date, the potential to extend applications of this technology to orphan crop improvement may be significant. Promising transgenic lines for about 20 different crops, including pepper, squash, and sweet potato, are currently being field tested by public research institutions in at least 10 developing countries. These lines, which have been engineered for particular traits like virus resistance and pest control, could provide significant benefits to poor farmers who cannot afford chemical controls for crop loss.

Genes used in direct ene transfer: A wide range of approaches employing transgenes is available for controlling herbivory, but by far the most extensively used method to date has been "Bt" technology Different Bt toxins are lethal to the immature or larval, herbivorous stages of lepidoptera, diptera, and coleoptera, many of which are major crop pests. Beyond the use of Bt, several additional insecticidal products are produced by plants, animals, and microbes that are also being evaluated for the control of herbivory. For example, many plants store large amounts of specialized proteins in their seeds, and some of these proteins have insecticidal properties.

Insects that consume plant seeds are particularly dependent on their ability to digest seed proteins using one or more proteases, and these inhibitors have been engineered into other plants to interfere with the feeding of insects on leaves and flowers. Protease inhibitors have also found to be effective against some of the plant pathogenic nematodes. A number of enzymes have also been tested for their possible use as insect resistance determinants. One of these enzyme groups is the chitinases, which target an essential structure of macromolecules (chitin) in insects and fungi. Three additional groups of enzymes have been designed to modify the chemistry of the plant in ways that make it less palatable as an insect food source.

Finally, in many crops, especially the grains, a major product stored in seeds is starch. Insect pests of these crops depend on their ability to digest starch using the enzymes - and -amylases. Some plants encode a gene for an inhibitor of - and -amylase; these genes have been used in a variety of cases to produce insect resistant plants. Transgenic approaches for reducing crop losses from disease have also been widely explored, with significant success in the area of plant viruses. Reports of success on transgenic technologies for controlling plant viruses abound in the literature and encompass numerous crops including rice, wheat, maize, barley, perennial rye, legumes, potato, sweet potato, several temperate vegetable species, and papaya.

Plant viruses represent a major biotic constraint on agricultural production; viral symptoms commonly include stunting, reduced yields, or death. Viruses are usually transmitted to plants via insect, beetle, nematode, or other types of vectors, and as a result, resistance is often the only effective form of viral control in agricultural systems. While genetic resistance exists in many crops, it is not always well understood or identified. For orphan crops in particular, even the most basic knowledge of resources for genetic resistance is typically lacking.

Reports from the mid-1980s confirmed that when portions of plant viral genomes, such as the viral coat protein gene, were transferred and expressed in a plant genome, resistance to the virus was observed. Since these reports were released, many types of sequences from both RNA and DNA plant viruses have been expressed in a diverse set of crops resulting in resistance to a wide array of pathogens. More recently, scientists have turned their attention to resistance resulting from expression of genes from other sources, including mammalian antibodies, anti-viral compounds, and sequences that act to silence viral genes or silence host genes necessary for viral infection.

To date, however, resistant varieties resulting from "second-generation" strategies are not yet in use. Nonetheless, future strategies to limit crop losses to viral disease will most likely encompass upstream research on sequences that silence plant genes and on direct gene transfer to move plant genes beyond the boundaries defined by sexual hybridization.

Genetic engineering for insect or virus resistances is a so-called "implementation technology", meaning that once the genes are available, they must be combined using an "enabling technology" required to move the genes into target crops. Thus, in principle the sequences used that encode the "pesticidal" activity, combined with appropriate regulatory sequences that work in the target crop, can be used in any crop plant. In this way, genes isolated originally for use in a major crop such as wheat or soybeans can be readily modified for use in orphan crops. The technical challenges for scientists applying these technologies to orphan crops are threefold.

The first challenge relates to the extent of knowledge about the sensitivity of the pathogens or pests of orphan crops to forms of control that have been useful in developed countries with temperate environments. Very little information on the usefulness of known transgenic approaches is yet available for many of the crops and the initial costs of developing this information can be significant for the lesser-studied orphan crops. For example, work in India on insect resistance for chickpea showed that the proteins encoded by the previously isolated proteinase inhibitor genes were at best marginally effective against insect pests of chickpea.

The researchers then turned to isolation of orthologous genes from crop plants such as Asian winged bean to find more effective proteinase inhibitors. The second challenge concerns the availability of techniques for transferring genes into orphan crops. In some cases, the transferability of gene transfer methods has been fairly easy and direct from model species to orphan crops. But in other cases, significant new and expensive work has been needed to modify-or to develop entirely new-transformation technologies for these crops. And finally, a major limitation of the use of these promising technologies in the orphan crops to date is the paucity of approaches that can be successfully applied to the control of diseases caused by bacteria, fungi, the oomycetes, and nematodes.

GENOMICS-GUIDED TRANSGENES

Using "genomics-guided transgenes" (GGT), native genes or homologous genes from closely related species that modify a plant's metabolism in a manner similar to natural or induced mutations are used for direct gene transfer. Dominant alleles that confer important agronomic traits of interest but that are scarce in breeding populations can be identified and inserted into breeding lines, based on knowledge gained from model gene studies.

An example of this approach is the use of the Xa21 gene, which provides resistance to the bacterial blight disease of rice. Xa21 was originally transferred from a wild relative to cultivated rice by conventional breeding. The gene has been moved among genotypes by MAS and has also been cloned and utilized through genetic engineering. GGT can comprise a wide diversity of genes, although many of the modified traits using this approach to date were already familiar, such as male sterility, seedless fruits, delayed spoilage, and dwarf stature.

BENEFITS OF ORPHAN CROPS

The most obvious example for the promising applications for orphan crops, is the grass family, or Gramineae, the most economically important plant family. There are around 10,000 graminaceous species, which include the world's most important cereal crops. Knowledge of the taxonomic structure facilitates the use of information from the more studied species to the orphans. There are five sub-families in the grass family. Rice is a member of the ehrhartoid sub-family and has the smallest genome. The pooids include wheat, barley and rye. The panicoids include maize, sorghum, and pearl millet, all of which are important staples in sub-Saharan Africa and other areas.

The chloridoids include the Ethiopian staple tef, as well as finger millet. The bambusoids include bamboo. Within the grass family, genomes vary substantially. While the genomes show similarities in gene organization, there is evidence of significant variation in genes, chromosomes and genomes. Based on a framework of restriction fragment polymorphism (RFLP) maps, the cereal chromosomes show gross synteny. At a finer level, however, there are many deviations from synteny, such that the predictive power between taxa ranges generally between 50 and 75%. Even between maize lines, the gene content of a given chromosomal region vary.

Some of the genes affecting important traits such as seed size, seed dispersal, and flowering time have been shown to exist in corresponding sites in the genomes of rice, sorghum and maize. Several QTLs conditioning resistance to the blast disease pathogen were found to exist in the same locations in both rice and barley. In many cases, therefore, knowing the location of a gene or group of genes affecting a key trait in one species will be useful for efficient identification of genes that are of interest for another trait.

If the actual genes are known, then the gene discovery process is less dependent on synteny, as the genes can be traced by sequence similarity. Similar examples are found for a range of plant families. For instance, the Solanaceae include many vegetable species that are extremely important as food and cash crops. For tomato, a substantial amount of genetic and genomic information is now available. The situation for potato and pepper is intermediate. The chromosomes of these species are syntenous, and the genes for various characters have been found to reside at syntenous locations. For various African indigenous vegetables (AIVs) in the Solanaceae, there is virtually no information available, but information from the better-studied solanaceous crops would doubtless be an asset if an investment were made in the AIVs. A similar situation exists for the legume family.

Despite identifiable niches, the task of implementing biotechnology approaches to germplasm improvement is not simple, particularly in poor regions. The process of analyzing the diversity of genes related to a trait of interest, and identifying and marking the useful alleles for selection, requires good access to current scientific literature and the internet, with at least modest bioinformatics capacity. The best promise for implementing a specific biotechnology strategy for germplasm development in orphan crops holds when research on a particular trait is pursued across a group of related species, one or more of which is not an orphan. The most rapid progress is likely to be made when there is a model crop in the same family as the orphan crop. Circumstances will vary greatly across crops and countries.

In the sweet potato case, weevils are known to reduce yields by 60%, and in some cases by 100%. Average yields, currently at 4.4 tons/ha, would thus rise to over 7 tons/ha with fully effective resistance. Viewing potential success more conservatively, a 25% increase in yields from weevil resistance would raise the average to 5.5 tons/ha. Given a sweet potato area of 570,000 ha in Uganda and a price of US$ 88/ton, the gain from achieving Bt-derived weevil resistance on half the area would be over US$ 27 million per year. If the technology could be distributed throughout the Sub-Saharan Africa on one-half of the sweet potato areas, the gross annual benefits would be about US$ 121 million.

Assume that twice the initial investment would be needed to move the resistant cultivars into the field in Uganda, and that five times the initial amount would be needed to disseminate the technology throughout the Sub-Saharan region. Even with these expenditures, the net payoffs would still be large. Moreover, because sweet potato is a vegetatively propagated crop, the biosafety risks of introducing Bt in this case would be much lower than for maize or cotton. Investing in marker assisted selection for blast resistance in finger millet could also prove successful within the next decade, particularly given the gains made in identifying genes and QTLs for blast resistance in rice.

Finger millet yields are typically reduced by 35% or more as result of neck and finger blast.Successful breeding for blast resistance could increase average yields from 1.3 ton/ha currently to over 1.75 ton/ha. A more conservative estimate of potential success would be to achieve a 15% gain in yields from resistance on one-half of the total area. Given an area of 3 million ha planted to finger millet in India, another 1 million ha in Africa,and a price of US$ 96/ton, the gross annual benefits would be more than US$ 38 million globally. Again, even if large costs were incurred to move the technology into the field, the net pay-offs would still be large.

Greater challenges exist with a crop like tef, where the technology path leading to the creation of dwarf varieties is more obscure. In all cases, however, there are benefits associated with the investments, including the enhancement of scientific capabilities and human capital development. The probability of success, the potential financial returns, the collateral benefits, and the constraints on technology adoption should all be weighed.

A technological continuum exists for germplasm improvement, which ranges from simple selection techniques, to conventional breeding approaches with farmer participation, to biotechnology developments in advanced laboratories. Unfortunately, the continuing decline in funds for practical crop improvement, particularly in the international public sector, results in a fair amount of investment in research, and a weak application to practical problems. Applying molecular techniques to basic research in plant biology has led to a much deeper understanding of the genetics of key traits, but not always to a clear idea of how plants can be improved.

Appropriate use of biotechnology depends on the agricultural problem at hand, the biological properties of the crop, and the economic and social infrastructure that supports crop research. The best chances for harnessing the gains from biotechnology exist when the science is integrated into breeding efforts, farm management, and seed production and distribution. Even with integration, the benefits from advanced science depend critically on the institutional, human capital, economic, and political context of the recipient countries. Useful generalizations about potential field-level applications of biotechnology for orphan crops are made difficult by the great diversity of crops and socioeconomic settings.

Nevertheless, orphan crop improvement is likely to benefit from attention in three areas: dispelling the myth that biotechnology is equivalent only to genetic engineering applications of Bt insect resistance and herbicide resistance; creating new incentives for public-private partnerships; and fostering new institutional arrangements that combine efforts across whole sets of major and orphan crops. Many people in the policy community-and in the general public-believe that agricultural biotechnology encompasses only transgenic approaches to insect and weed resistance. This focus on GMOs has pitted consumers against industry, Europe against the US, and environmentalists against each other as they debate the relative costs of pesticide use and genetic engineering.

While the debates are valid in their own right, they overlook the fact that many new genetics and genomics techniques do not involve the insertion of foreign genes into plants. The use of molecular tools to understand the genetic basis of crop traits for indirect selection and breeding is beginning to offer an alternative to traditional transgenics. These technologies will help alleviate some of the constraints surrounding intellectual property rights and biosafety that limit the dissemination of GMOs in developing countries.

Many developing countries have been reticent to employ GMOs, lest their use interfere with trade and aid. It is the responsibility of agricultural policy analysts and practitioners to help redirect policy and public discussions on the future of biotechnology. The debate should focus foremost on the needs of poor farmers and consumers, and the tradeoffs they face. To the extent that conventional breeding can be facilitated by advanced genetics techniques, biotechnology should be encouraged as one tool in the agricultural development package.

For major crops, such as wheat and rice, much of the linking of research with farmers has been performed by Centers of the CGIAR. Limited inroads on millets, cassava, bananas, and a few other crops have been made by additional CGIAR Centers, but there are large numbers of orphan crops where virtually nothing has been done. It is on this point that the new AATF instituted by the Rockefeller Foundation holds considerable promise. Its mission is to help design the relevant templates, protocols, and procedures that will lower the transaction costs of applying biotechnology to major and orphan crops in Africa. By doing so, it will provide a model for other regions.

CHAPTER 10

AGRICULTURAL AND SUSTAINABLE DEVELOPMENT

The term sustainable development is used in a broad sense to include poverty alleviation, food security, enhanced health and environmental care systems, promotion of industrialization and economic productivity. Emphasis is placed on the applications of recent advances, rather than details of the underlying science itself, in order to portray the potential benefits and concerns associated with current practices in agricultural biotechnology.

According to the World Bank, approximately 1.2 billion people globally are absolutely poor, living on less than US$1 per day. Nearly twice this number live on less than US$2 per day. Environmental degradation, hunger, malnutrition, mass starvation, disease and death are phenomena regularly experienced by millions of people in Africa. Estimates released by the Food and Agriculture Organization (FAO) of the United Nations in November 2003 state that there are 842 million poor and chronically undernourished people worldwide.

The majority of these people barely survive on less than a dollar a day. Out of the 650 million Africans, 250 million are chronically undernourished, over 40 million children are severely underweight, 50 million suffer from vitamin A deficiency, while 65% of women of child bearing age are anemic. About 90% of

Africa's poor are rural, resource-poor farmers, their families, and the landless poor who depend on agriculture for their livelihoods.

The rest are mostly urban, jobless slum dwellers. Women produce 70 to 80% of the food in Africa compared with 65% in Asia and 45 % in Latin America and the Caribbean. Between 1970s and 1990s, per capita food production in Africa declined at a rate of 2% per year. Despite this trend, population growth continues at a rate of 2.8% annually, against declining economic growth and reduced investments in agriculture.

The natural resource base of suitable land, water, forests and biodiversity which largely determines the potential of agriculture, are under severe threat. It is estimated that 80% of the land in Africa is threatened by degradation, while 2 million hectares of forest are lost annually. Water and wind erosion, loss of soil nutrients, overgrazing, salinization, acidification, pollution, compaction and waterlogging, are some of the factors that lead to environmental degradation, especially land degradation and deforestation.

In order to obtain solutions to these problems, and to understand how biotechnological tools may be applied, it is useful to track past events or "revolutions" that appear to have acted as vital catalysts in helping other countries elsewhere to achieve food selfsufficiency, especially the Green Revolution in Asia.

GREEN REVOLUTION

Over 200 years ago, during the early decades of the industrial revolution, the English economist Robert Thomas Malthus suggested that human population growth would outstrip food production since population increases faster than the means of subsistence. Although this may be partly true, the discovery of energy sources, notably coal and natural gas, spawned an "energy revolution" that impacted positively on the means of agricultural production in a way that enabled food production to match or even potentially outstrip population upsurges in Europe and North America.

The ability to manufacture nitrogen and other fertilizers ("fertilizer revolution"), and to mechanize agriculture, especially through use of irrigation ("water or irrigation revolution") has allowed countries such as Israel to become self sufficient in food production in otherwise hostile environments. Whereas about 40% of the global population did not have adequate food supplies in the 1960s, approximately 20% of the current global population does not have enough food in 2003.

However, the number of people lacking adequate food remains the same, and this number is largely composed of the poor in sub-Saharan Africa. The term Green Revolution refers mainly to the large increases in grain production achieved by certain underdeveloped areas, particularly India, Pakistan and the Philippines from the late 1960s. This fete was accomplished through the use of high yielding hybrids, chemical fertilizers, new crop management strategies and harvesting methods.

Although questions have been raised about the environmental "sustainability" (e.g. soil integrity) of Green Revolution practices, all are agreed that countries that successfully underwent this Revolution changed from being net food importers to net exporters of grain to date. But can the success of the Green Revolution be repeated on a global scale, in many diverse agroecosystems, and in an equitable, sustainable and environmentally friendly manner.

If such a "Doubly Green Revolution" is possible, how shall we ensure that SSA, which needs it most, does not miss out. And for the sake of this discussion, what role will biotechnology and genomics play. From the very outset, one thing seems certain - that the gene revolution is only a part, albeit invaluable, of a holistic, integrated, multidisciplinary approach needed to address the constraints highlighted above. The salient scientific research advances and trends in plant and animal agricultural genomics and the implications of these advances for sustainable development in SSA. Tissue culture-related technologies that do not involve direct DNA manipulations, such as micropropagation, haploidy and embryo culture, are not covered.

AGRICULTURAL GENE REVOLUTION

The genome of any organism represents the sum of all the genes in that organism. Structural genomics is concerned with the elucidation of the nucleotide base composition of gene sequences, whereas functional genomics focuses on how genes are regulated and expressed, at the level of DNA, RNA (trancriptome) and protein (proteome). The aim of genomics is to discover genes that code for Mendelian and complex traits, and to utilize the knowledge gained from such studies to address problems in fundamental (e.g. biochemical and developmental biology) and applied biology (e.g. agricultural, industrial, medical and environmental applications) and product development.

However, there is no fine line between basic and applied genomics and proteomics research since most basic research often has direct bearing on applied sciences. For example, basic biochemical studies of oil, protein, starch, cellulose synthesis and secondary metabolism, or research in developmental biology of fruit ripening, fertility, seed development and growth characteristics have obvious applications in crop agriculture and forestry at the farm, field and post-harvest storage levels.

Large-scale Genome

Large scale DNA sequencing can be performed on genomic DNA or on complementary DNA (cDNA). Using these strategies, the genomes of many organisms have now been either fully or partially sequenced and resultant sequences deposited in nucleotide or protein data banks, many of which are freely accessible on the internet. One notable achievement was the landmark completion of the human genome sequence in 2001, in addition to several representative microbial, insect, animal and plant genomes.

The completed human genome is expected to facilitate identification of all genes that contribute to human disease. In the drug industry, many pharmaceutical companies are valued at hundreds of billions of dollars, much of which depend on their ability to exploit genomic information in developing blockbuster drugs (pharmacogenomics). In animal agricultural genomics, advances in the use of cloned and transgenic animals in research programs will have a profound impact in many areas such as development of new diagnostics and vaccines and enhanced growth characteristics and reproductive performance.

Another key area will be understanding the genomes of microorganisms that are important to the productivity and sustainability of agriculture and forestry, and to the safety and quality of the world's food supply chain. High volumes of genome data is accompanied by an even much greater increase in complexity of such genome data, thus creating the need to appreciate both the scale of the challenge of large genome analysis and the limitations of current gene prediction methods and understanding. Another feature of large-scale sequencing projects is that they are very expensive and often require forging partnerships from the private and public sector, and may involve different countries, as is the case for the Human and Arabidopsis Genome Projects.

Genome Mapping

Humans have genetically modified plants and animals for as long as they have cultivated crops or kept livestock. This has heppenned either inadvertently or through structured breeding programmes. Traditional plant and animal breeding uses phenotype (e.g. crop yield, or milk production) to predict and select genotype, i.e, specific crop variety or animal breed. The phenotype is the product of distinct genotypic and environmental effects. Phenotypic selection based on traits that are conditioned by additive allelic effects has now been overtaken by marker-assisted selection (MAS).

MAS relies on the existence of heritable genetic or molecular (usually DNA) markers that are associated with economically important traits. The predictive value of genetic markers used in MAS depends on their inherent repeatability, map position, and linkage with economically important traits [quantitative or qualitative]. The major advantage of MAS is that it provides a potential for increasing selection efficiency by allowing for earlier selection and reducing population size used during selection. Not only is the efficiency and precision of breeding enhanced, but costs are also reduced.

MAS of many important food and cash crops, trees and important farm animals now allows new cultivars and breeds to be developed far more rapidly than the "hit and miss" approach typical of conventional selective breeding programmes. In addition to genome data, the physical maps of many crops, pathogenic microorganisms and some livestock are now available, for example for corn, rice, some viral, bacterial and fungal species and cattle. Efficient use of genome data from such organisms offers unprecedented potentials for crop and animal improvement. Among cereal grasses, there is very high synteny, in terms of gene structure and order, e.g. between rice, wheat, corn, millets and sorghum.

By taking advantage of comparative genomics, it should be possible to establish genetic and physical maps of socalled African orphan crops such as tef, sorghum and millets. The extensive seed banks of these African crops can be used to initiate large-scale breeding programmes that utilize molecular markers to accelerate the evaluation of the outcome of the crosses. One example is the accumulation of several desirable traits (also called pyramiding) in the same genotype.

Another good example is marker-assisted backcrossing (MABC) programmes which utilize mapped markers to select backcross progeny that have the highest percentages of the recurrent parent genome and the minimum number of donor parent segments. By using new MAS strategies and software tools for precise genetic mapping, high-resolution chromosome haplotyping and extensive phenotyping, it is now possible to create novel, superior crop varieties and livestock breeds in what Peleman and van der Voort have termed "breeding by design".

However, this will require prior understanding of the molecular genetic control of agronomically important traits and the allelic variation that exists at loci controlling those traits, before the molecular breeder is able to design superior genotypes "in silico". Lastly, it should be mentioned that most of the technologies being developed for MAS are also applicable in the evaluation, conservation and use of biodiversity, including microbial, animal and plant genetic resources.

Animal Agricultural Genomics

Cloning and transgenic animal research has attracted a lot of attention, particularly since the cloning of Dolly the sheep. Although human cloning is banned, the use of cloned and transgenic animals and insects is now being done routinely in several laboratories around the world. Use of cloned animals may be useful in creating "herds" with uniform product characteristics. So far, the area that already has had impact is the use of transgenic animals such as chickens and cattle to produce pharmaceutical or therapeutic proteins in eggs and milk.

Though difficult to predict with certainty, the overall global impact of transgenic and cloning technologies in animal agriculture is expected to be low, compared with crop agriculture and forestry. In animals, the complexity of multiple organs, tissues, and cell types, combined with mechanisms of crosscommunication complicate all studies aimed at improvement. Another problem is the lower degree of totipotency exhibited by animal cells, making regeneration scientifically harder and more expensive to perform. In addition, ethical and religious considerations pertaining to any form of genetic tampering are more rigorously pursued with animal and human subjects than with other organisms.

However, the greatest impact of genomics is being realised in the development of new diagnostics and vaccines for animal diseases and in the production and use of animal feeds. Many vaccines and growth hormones are amenable to recombinant DNA manipulations, hence allowing avenues for development of animals, especially livestock breeds, with enhanced growth characteristics. For resource-poor farmers in Africa, utilization of improved disease control strategies through use of disease resistant livestock and improved diagnostics, vaccines and therapeutics for disease prevention and treatment would greatly help secure livestock assets, with the greatest impact likely to be felt among pastoralist communities in the arid and semi-arid regions.

In trying to achieve these objectives, it is crucial to target local genetic resources rather than exotic livestock genotypes as sources of useful genes, as illustrated by presence of disease resistance genes in locally adapted West African dwarf goats and Maasai sheep. Use of genomics tools to address these constraints would increase productivity, farmers' income and food security, while also reducing chemical and drug usage and hence improving environmental health and sustainability.

Production of transgenic animals with improved traits such as enhanced reproductive performance, disease resistance and tolerance to environmental stresses is still not a reality, although knowledge about the genetic and biochemical control of such traits make current research towards achieving these goals realistic and achievable in the long term. As with the Green Revolution, there has been an equally powerful revolution in the field of animal feed efficiency and nutritional requirements arising from advances in animal genetics. For example, it now takes less than 1 pound of feed to produce one pound of weight gain in broiler chickens, whereas it took 6 pounds of feed to produce the same increase in weight 100 years ago.

These improvements in animal feed have been made at the level of both macronutrients and micronutrients or as pharmaceuticals. Macronutrients, namely carbohydrates, lipids and proteins (amino acids) constitute the bulk of feedstuffs, and are primarily used as energy sources. Micronutrients (e.g. minerals and vitamins) are organic or inorganic compounds present in small amounts and essential for good health. Increasingly, bioproduction of these substances is through recombinant DNA tools utilising the latest knowledge in both structural and functional genomics.

Examples include dietary supplements, such as vitamin D and E (antioxidant), amino acids (e.g. methionine to improve immunity), enzymes (e.g. to improve digestion and remove antinutritional metabolites and toxins), antibiotics and growth hormones (to improve health and enhance growth rate). Other advances include use of prebiotics and probiotics to improve gut microflora. In the developed countries, a very significant proportion of crops, especially corn, are grown for animal feed in the form of hay, silage, forage or grain. In these countries, genetic modification of crops to enhance feed value has reached farmers.

Such crops include silage corn with reduced lignin content to enhance digestibility, oil rich corn and corn with enhanced protein content. Crops having reduced non-digestible components such as the oligosaccharides stachyose, raffinose and galactose, or with enhanced resistance to fungal pathogens, or a combination of these traits, would certainly be welcome, especially to overcome contamination of feeds by mycotoxins. To date, many animal gene sequences, including over 250,000 cattle genes and more than 150,000 pig genes are available in public databases such as the Institute for Genomic Research.

In the coming decades, the vast amount of information gained from functional genomics studies will help to: 1), target relevant genes for drug studies; 2), develop effective therapies for disease treatment; 3), determine what hormones and metabolites are involved in animal responses to environmental and developmental cues; 4), determine true impacts of husbandry practices on animal health; 5), determine true impact of genetic selection on animal health; 6), find harmful and beneficial forms of genes for genetic selection, and; 7), guide strategies for transgenic research to improve animal health.

These examples provide sufficient evidence to spur future investments in animal genomics to address areas related to production, product quality, food and feed safety and health. Uptake of these new technologies in Africa is imperative not only for solving constraints but also for helping to diversify animal agriculture while at the same time promoting croplivestock integration as a way forward for uplifting rural economies.

Plant Genomics

In addition to marker-assisted breeding discussed above, identification of the genes conferring the agronomic traits to our crop and forest plants and engineering of novel traits will be the base of a sustainable agriculture, forestry, industry and environment care. Methods of gene identification include direct sequencing of genomic DNA and ESTs, gene disruption and trapping strategies, and differential display. Furthermore, several methods to analyze RNA, proteins (proteomics) and metabolic pathways (metabolomics) at a large scale are now routine and continuously being improved.

Use of highthroughput technologies such as microarray and gene chips for expression profiling and improved tools of analytical organic chemistry allow the establishment of profiles of all metabolites present in specific plant cells and tissues. The later approaches will allow unravelling of biochemical pathways, identification of genes encoding biosynthetic enzymes and production of secondary and intermediar metabolites in crop plants. Given the ever-increasing knowledge and information pertaining to genomics, new bioinformatics and computational biology tools are needed to analyze, interpret and utilize the enormous body of data and knowledge generated.

With major breakthroughs in genomics and analytical chemistry, plant biologists are now aiming to integrate transcriptomic, proteomic and metabolomic datasets using bioinformatics tools in sufficient detail that affords system-wide predictions of plant development in response to both genetic and environmental perturbations. This new era of "systems biology" is probably the ultimate undertaking in terms of attempting to manipulate or improve plant traits for agricultural uses in an environmentally sustainable manner. One key trait that has defied scientific unravelling is the phenomenon of heterosis.

A systems biological approach to define how plant genomes interact to create phenotype is needed to arrive at a final resolution of this phenomenon. Typically, target traits for genetic improvement fall into three classes, namely, input (agronomic) traits, output (nutritional quality) traits and physical traits. Agronomic traits are both biotic and abiotic (e.g. better adaptation to heat, drought, salinity, acidity, heavy metals, water logging and nutrient [especially nitrogen and phosphorus] availability). Output traits include content and quality of starch, protein, oil, nutritional elements.

Physical traits include plant architecture, shelf life and podshattering. In future, the most challenging studies with immediate impact on crop and tree agriculture will include regeneration (totipotency) and transformation capacity, vegetative seed production (apomixis), seed molecular biology, haploid plants; heterosis or hybrid vigour, symbiotic nitrogen fixation, parthenocarpy, plant architecture, flowering and drought resistance. For example, many genes controlling plant architecture and flowering time, and which have potential utility in crop improvement in Africa, have been characterized.

Some of these genes can be either up- or down-regulated in transgenic plants to dramatically alter such properties as stem strength and length, leaf size and shape, the form and polarity of organs such as fruits and flowers, the activity of various meristems, root branching patterns, the cell division cycle that plays a big role in determining the overall rate of plant development and genes controlling circadian rhythms, flowering time, and vernalization.

As an example, introduction of one or several possible dwarfing genes to the tall East African Highland banana may reduce the size of this important fruit crop. Also, over-expression of gene(s) that can shorten the cell division cycle could speed up the otherwise very long growth cycle of this crop. New technologies and strategies that rely on mechanisms such as transcriptional and posttranscriptional gene silencing are increasingly being used to address economic crop constraints in ways that were hitherto not contemplated.

One such technique involves the silencing or down-regulation of specific genes by double stranded RNA (dsRNA) through a technology referred to as RNA-mediated interference or RNAi. In some cases, it has been shown that the presence of just a few molecules of dsRNA is sufficient to knock-down the expression of a gene that is homologous to the dsRNA. This approach opens up many avenues for utilizing already large amounts of specific genomic and cDNA sequences encoding known functions, in agricultural genomics with respect to priority constraints of economic value.

Of particular importance to Africa will be manipulation of genes involved in abiotic stress mechanisms, especially drought stress and water use efficiency. It is now becoming clear that gene expression associated with dehydration (and freezing) stress involves both abscisic acid (ABA)-dependent and ABA-independent signal transduction pathways. Overexpression of ABA induced genes is thought to result in enhanced stomatal responsiveness leading to closure of stomates (and henced enhanced "staygreen" phenotype) during water stress.

Presently, plants engineered with genes that encode enzymes that synthesize osmoprotectants or genes that encode transcription factors such as the CBF/DREB family, provide the most promising economically viable technology to have applicability beyond the "prove of concept" phase. Mendel Biotechnology (California) is currently licensing WeatherGard™ genes which are regulatory (CBF/DREB) genes that allow plants to be engineered to be more tolerant to drought.

This is an opportune time to test these technologies in the farmers's fields through genetic transformation. Many research groups have also isolated genes from dessication tolerant plants such as the resurrection plant in South Africa, a desert moss that can slow its activity down to zero when water is scarce, and revive from almost complete dehydration. ABA induced genes have also been identified from the model moss Physcomitrella patens through a pilot EST sequencing project.

Other strategies include manipulation of specific genes that control the pattern of root branching, to create deeper, perhaps more branched roots to increase the surface area for water absorption, and/or in combination with use of desiccation stress-inducible promoters. Directly, or indirectly, plants provide all of humanity's food, most of which comes from only a few crops, namely rice, maize and wheat. Over many years of crop agriculture, the focus has been on agronomic traits such as yield, productivity and pest resistance, while nutrient content has largely been ignored, with the result that most food staples are severely limited in nutritional composition.

In Africa, 250 million out of 650 million Africans are chronically undernourished, over 40 million children are severely underweight and 50 million suffer from vitamin A deficiency, while 65% of women of child bearing age are anemic. To alleviate these problems, research is underway to fortify food staples to enhance levels of essential and non-essential micronutrients and macronutrients, such as vitamins, minerals and proteins.

Examples include cereals and sweetpotato with enhanced levels of vitamins and/or proteins in their seeds and tubers, respectively. This field of nutritional genomics takes advantage of the many genes that have been cloned for vitamin pathways and for the synthesis of many other "nonessential" compounds and macronutrients. In future, it should possible to directly manipulate the content and composition of many nutrients in staple African food crops such as cassava, sweetpotato, banana, cowpea, maize, millets and sorghum.

Other benefits will include lowering the levels of antinutrients (e.g. phytic acid), toxins (e.g. cyanide in cassava) and allergens in seeds, food grains, leaves and tubers. Transgenic plant production of edible vaccines and antibodies (plantibodies), as well as products of therapeutic, pharmaceutical and industrial value, is also possible. Products with industrial uses include latex, oils (palm, soja, rape seed), speciality oils and waxes, sugars, starch, cellulose, wood, monomers for biodegradable plastics, etc. Although this field of "molecular farming" is still a long way to go, it has the potential to spur rural agroindustrialization and enhance crop diversification for new uses, and hence to increase sources of income for resource poor-farmers who depend on agriculture for their livelihoods.

Public Concerns

Public concerns about bioengineering fall within five categories:

— safety issues related to human health, food safety and antibiotic markers

— concerns about detrimental environmental impacts such as loss of biodiversity, gene transfer to wild relatives, non-target effects arising from insect resistant crops and escape of bioengineered crops

— regulatory concerns, such as unsatisfactory regulation, labelling policy and the right to choose

— economic concerns, such as control of biotechnologies by big multinational companies and rich nations resulting in limited impact on developing countries, especially on small and medium family farmers in Africa

— ethical concerns.

Some of these concerns are legitimate, others clearly result from lack of information and/or misinformation about the technology, whereas others are deliberately not genuine. But however the case may be, these concerns largely determine whether or not biotechnology is accepted and/or adopted by end-users, mainly farmers, industry and consumers. Indeed the precision of recombinant DNA technologies and our current knowledge about gene regulation, makes these technologies much more predictable than random mixing of genes that occurs through conventional breeding.

Data from first generation transgenic crops shows significant economic and environmental benefits, such as increased yields, easier weed control, reduced pesticide use on insect tolerant crops and reduced soil erosion resulting from non-tillage practices when growing herbicide tolerant crops. However, it must be noted that no technology is riskfree, hence the concerns mentioned above need to addressed through sound science rather than mere emotion and unfounded speculation. Furthermore, it is important that these concerns are also raised and addressed with respect to classically bred crops and agricultural practices.

One reason for doing this is because we know that many conventionally bred crops such as tomatoes, cassava, broccoli and potato contain unsafe levels of toxins, allergens and/or other antinutrients that have been ignored by those who claim that bioengineered foods which have undergone and passed rigorous food safety tests are (still) not safe to eat. Responsible use of modern biotechnology must also take into account societal (including ethical and religous) values, trust, familiarity, controlled risk and tangible benefits to the consumer.

So far, scientists have mostly followed scientific rationality and thereby failed to understand that public opinion follows not only science but also other "non-scientific" considerations. It is therefore imperative to educate and make the public aware of the importance of science in decision making by moving from "educating the public" to engaging with the public, through discussions with stakeholders such as farmer, consumer and regulatory organisations.

In all this, the consequences of NOT applying modern biotechnology tools in agriculture must be explained. In Africa, the fruits of biotechnology are still far off, and yet we seem to be in blind danger of terminating this technology before it even germinates.

CHAPTER 11

COMPLEX TRAIT GENETICS

While genetics provides the scientific basis for the breeding processes we use today, for the majority of the history of applied breeding the concept of a gene was unknown. Selection was conducted on the phenotypes of individuals in ways that was a mimic of the Darwinian process of natural selection, which was itself an undefined concept prior to 1859. Further, for the majority of the 20th Century the genes underlying quantitative traits were at best theoretical constructs and were not directly viewed or manipulated in breeding programs to bring about changes in phenotype. It was only in approximately the last quarter of the 20th Century that we have observed technological advances that provided the opportunities to study genetic variation at the DNA sequence level. These technologies have enabled us to go from the phenotype to the gene, and in some cases from the gene to the trait phenotype.

Many attempts have been made and are currently underway to construct relevant gene-to-phenotype (GP) models for traits to assist the plant breeding process. The approaches being used are diverse and in many cases unproven. This is a time of exploration of many novel ideas on how to approach the GP problem. Quantitative genetic theory, with all of its assumptions, was founded with the goal to understand the genetic basis of the variation for quantitative traits and to use this knowledge to make predictions about the properties of genes in populations of genotypes and the outcomes of artificial selection and evolutionary processes.

Despite the significant advances we have made in the range of molecular technologies available to study the genetic architecture of traits, the GP prediction problem still exists today as a major challenge for most of the important traits in plant breeding. If anything, today the magnitude and complexity of the task involved in predicting phenotypic variation based on knowledge of DNA sequence variation is now more obvious to a wider audience than was previously the case.

Here we can see the components of a quantitative framework, that was done by some researchers. It can be used to investigate some expected properties of traits under the influence of selection for the continuum of simple to complex traits. Classical quantitative genetics, in combination with linear statistical models, provides the conventional approach to prediction of expected response to selection. This framework is differs from much of the classical theoretical framework of quantitative genetics in that it is implemented and the predictions are obtained through computer simulation rather than through seeking solutions to linear statistical models using approaches from calculus.

The motivation for developing and using a simulation framework is the difficulty of extending the classical statistical framework to accommodate recognized properties of complex traits, particularly effects attributed to features such as gene-by-gene interactions (epistasis) and gene-by-environment interactions. The need to accommodate these properties of gene action in many GP models is indicated by experimental investigations into the molecular basis of regulation of gene expression, signal transduction pathways and other features of genetic variation at the DNA sequence level and their influences on phenotypic variation of traits within an organism that is undergoing growth and development within an environmental context.

The use of computer simulation to study problems in genetics is not new. However, many of the early applications of simulation were based on scaling up the classical simple Mendelian inheritance models, assuming independence of gene effects from the genes at other loci and of environment. Thus, many of the complex interactions we now consider as components of the GP models for some of the important traits were not included as features of the early simulation experiments. This limitation is not unique to the early genetic simulation experiments.

Many of the applications of simulation to problems in quantitative genetics today still only consider genes as independent Mendelian factors. Such limited treatment of the continuum of GP models is unnecessary today and is not recommended. Such approaches will give an overly optimistic and simplistic view of the expected outcomes of conventional and molecular breeding strategies.

E(NK) MODEL

The E(NK) model is an extension of the NK gene network model that was introduced and used by Kauffman to study the behavior of gene networks and their influences on organism development and evolutionary processes. Here the application of the E(NK) model is to the study of issues as they are relevant to plant breeding processes. The E(NK) model allows for the property that the influence of a gene network on determination of a trait phenotype can differ among environmental conditions. Thus, E identifies different environment-types within the context of a defined target population of environments, N identifies the different genes and K identifies the degree of connection between subsets of the total set of N genes, i.e. the gene network topology.

Thus, in the terminology of quantitative genetics the E(NK) model is a finite locus polygenic model that can be defined to include effects of epistasis and gene-by-environment interactions.

The parentheses around the NK term are used to indicate that the N genes can interact in different K ways to determine the trait phenotype in different E environment-types. To date we have used the E(NK) model as an organizing framework for the design of large scale computer simulation experiments that are conducted to investigate both the properties of GP relationships for a continuum of simple to complex genetic models and the power of different plant breeding strategies to achieve response to selection along this complexity continuum.

Kauffman gave a detailed explanation of the background and specification of the NK model. The background to the E(NK) genetic model and descriptions of its application to specific traits within a plant breeding context have been given elsewhere. We consider the use of "hybrid models", where the genetic component of the model is based on a combination of "explained" genetic variation attributed to defined genes (or Quantitative Trait Loci; QTL) and "unexplained" background genetic variation. The explained component of the genetic variation may be defined as an outcome of investigations into the inheritance of a trait by use of suitable experimental methods, such as inheritance studies using a genetic or sequence based map of the genome.

A typical experiment involves testing a sample of genotypes from a reference mapping population in a sample of environments. The phenotype for the *kth* observation on the *ith* genotype in the *jth* environment (P_{ijk}) can be described by the linear equation:

$$P_{ijk} = E_j + G_i + (GE)_{ij} + \varepsilon_{ijk} \quad (1)$$

where, E_j is an environmental effect, G_i is a genotypic main-effect, $(GE)_{ij}$ is a genotype-by-environment interaction effect, and ε_{ijk} is a residual effect. Linear statistical models, of the form given in equation (1), can be used to analyze means and variances associated with the genotype-environment system based on the experimental sample. It should be recognized that equation (1) is not a GP model; it is a statistical partition of inter-individual genotypic variation.

For a GP model we need a specification of the trait performance as an outcome of the combined effects of the N genes influencing the trait. If we use γ_n to refer to gene n, where $n = 1,\ldots, N$, and $\gamma_{g|n}$ to represent the specific g genotypic combinations of alleles for gene n, then any individual, represented as a multi-genic genotype G_i, can be considered to have a value in environment E_j from the combined effects of the N genes in that environment. We identify the combined multi-genic effects of the N genes for genotype i in environment j by $\Gamma_{(g|N)ij}$, and rewrite equation (1) as a GP model in the form:

$$P_{ijk} = \Gamma_{(g|N)ij} + \varepsilon_{ijk} \tag{2}$$

We use equation (2) as a compact statement that defines the phenotypic value for a trait measurement k on a genotype i in an environment j as the genotypic value, that is the outcome of the combined effects of the allele combinations for the N genes, and a noise parameter ε_{ijk} that can be interpreted as a random environmental and/or measurement error effect. From the above, the $E(NK)$ model is an alternative form of the genotypic component of equation (2) that identifies the combined effects of the N genes by the NK term and the specificity of the resulting NK genotype values to an environment by nesting the $(NK)_i$ term for genotype i within a defined environment E_j. Thus, we can rewrite the $E(NK)$ model in a form similar to equation (2) as:

$$P_{ijk} = E_j(NK)_i + \varepsilon_{ijk} \tag{3}$$

Thus, equations (2) and (3) define the rudimentary framework for computing the genotypic and phenotypic values for *i* genotypes in *j* environments from knowledge of how the *N* genes act to determine the trait phenotype. More generally we can recognize that the *E(NK)* model is itself a special case of the general expression for the equations of state of a system given by Casti:

$$y = \Phi_{\alpha}(u) \tag{4}$$

where; *y* is the phenotype, Φ is a mathematical relationship expressing the relationships among the observables, α is a parameter vector representing the genetic component of a system and *u* represents environmental conditions. Recognizing this relationship identifies and opens up a wide array of possibilities for studying the GP problem state space that go beyond the conventional framework of quantitative genetics.

PREDICTING PHENOTYPE FROM GENOTYPE

If we seek to improve the effectiveness of the conventional plant breeding process by using molecular technologies to design a knowledge-based approach to plant breeding, equations (2) and (3) emphasize that one of the major challenges that we face is how to discover the *N* genes that are important in determining the extant phenotypic variation for a trait and how to understand the various functions of these genes. Functional genomics technologies can then be applied in combination with appropriately designed genetic experiments to provide the basis for diagnosing the interactions among the genes and the construction of gene network knowledge.

A model based on the integration of such knowledge would then provide a starting point for understanding our current capacity to predict changes in the phenotype from directed changes in the genotype. Two approaches we can take to examine the magnitude of this GP prediction problem are:

1) Undertake an extensive experimental program to discover the genes and understand their function and role in determining phenotypic variation for the target traits, an activity that is currently underway in many research groups. Then use this knowledge to make predictions and conduct validation experiments to test the predictions.

2) Use the E(NK) model to create a simulated ensemble of many different "plausible" GP models and examine the robustness of alternative breeding strategies across these different GP models. We can combine both approaches by using the available experimental data to place the empirical GP models into the theoretical GP problem space.

From the body of quantitative trait mapping literature available today we observe that for every study we will obtain a statistical model that is a partial representation of the genotypic components of the phenotypic variation. Thus, we can rewrite a qualitative version of equation (1) to represent this situation as:

$$P_{ijk} = E_j + G_{Expl} + G_{UnExpl} + (GE)_{Expl} + (GE)_{UnExpl} + \varepsilon_{ijk} \quad (5)$$

where the subscripts Expl and UnExpl identify explained (or estimated) and unexplained components of the G and $G \times E$ components of the linear model, respectively. Here, we treat the explained and unexplained components of the model as independent terms. It is emphasized that this independence is not necessary and is only a subset of the more general case. If the explained component interacts with the unexplained component we have a much more complex situation than is indicated by equation (5). In the presence of such interactions we may expect to overestimate the level of predictability within the system based on the GP knowledge we have acquired. A tangible example of this situation would be when we move alleles of genes from one population into another and observe different effects of these alleles in these different contexts. Because of space constraints we do not give a formal treatment of the more general case here.

We can quantify equation (5) by replacing the explained component directly with the current version of the GP model for a trait. For example, if we had identified two QTL for the trait and QTL 1 is constitutive in its behavior and effects across environments whereas QTL 2 is facultative, with effects on the trait phenotype that are specific to the types of environment, using the notation above, but using Q for a QTL rather than γ for a gene, we could write equation (5) as:

$$P_{ijk} = E_j + Q_{(g1)i} + Q_{(g2)ij} + G_{UnExpl|i} + (GE)_{UnExpl|ij} + \varepsilon_{ijk} \quad (6)$$

where the estimates of the 2 QTL effects represent an explained component. While we can use estimates of QTL effects as a parameterization of the explained component of the model, this opens the question of how to appropriately represent the unexplained component. A classical approach would be to assume that the unexplained component was independent of the explained component and can be represented by effects drawn from some underlying distribution, such as the normal distribution. However, if we apply the same general arguments regarding the limitations of the assumptions of independent gene effects given above to the currently unexplained component we may consider that representing the unknown component of the model in such a manner may be an overly optimistic representation. An alternative approach that we have considered is to represent the known component by the QTLs and the unknown component by an ensemble of *E(NK)* model parameterizations. Thus, we can rewrite equation (6) as:

$$P_{ijk} = E_j + Q_{(g1)i} + Q_{(g2)ij} + E_j(NK)_{ij} + \varepsilon_{ijk} \quad (7)$$

where we have retained the two QTL from equation (6) as the explained genetic component and defined the unexplained genetic component as an *E(NK)* model parameterization. A macro-environmental effect E_j and a micro-environmental noise effect ε_{ijk} associated with measurement *k* on genotype *i* in macro-environment *j* are both retained in this model expression. The E_j term could in fact be omitted but we leave it in here for consistency and we define the macro-environments to be the same as the environment-types specified within the $E_j(NK)_{ij}$ component of the model. Equation (7) is an expansion of equation (3) using empirical results from the currently available GP model.

Equation (7) can be parameterized by specifying the effects for a subset of the *N* genes defined in equation (3) based on experimental estimates of their effects and specifying the remainder of the *N* genes by sampling effects from some underlying distribution. We refer to equation (7) as a hybrid model, where the explained component is parameterized by estimates of gene effects obtained from experimental results and the unexplained component is stochastically parameterized by drawing effects from a specified distribution of effects. Equations (5) - (7) represent one example of how this approach can be implemented within the framework indicated by equation (3). We can indicate a more general application for QTL models by writing:

$$P_{ijk} = [E_j(NK)_{ij}] + [E_j(NK)_{ij}] + \varepsilon_{ijk} \quad (8)$$

Equation (8) can be applied to incorporate effects attributed to epistasis in the form of QTL×QTL interactions in the explained or unexplained component. Further extensions of this framework that we have considered include allowing some of the *K* interactions to occur between some of the N genes in the explained component and those in the unexplained component to simulate the effects of partial explanation of a gene network or the gene-by-background effects often experienced in plant breeding, where QTL or gene effects are population (cross) specific. We can expect to observe a range of simple to complex situations and a mixture of successes and failures in breeding programs as we attempt to validate QTL effects and use the QTL alleles in applied breeding.

To study what we perceive to be a complex biological problem it seems appropriate at this time, while we are currently building the research foundations for molecular approaches to plant breeding, to explore some of the developments and emerging concepts in the area of complexity science and its potential applications to the study of complex biological systems. Casti gives an introduction to this field and some of the quantitative modeling tools that have been applied. Many of the applications of these methods have focused on the GP modeling problem. In general, but with a few notable exceptions, this work has tended to remain outside of the mainstream genomics, quantitative genetics and plant breeding communities.

The current separation of these efforts may in part be explained by a combination of the recent emergence of a critical mass of research in the complexity science field, the historical momentum behind the classical quantitative genetics approaches and the recent widespread availability of high throughput technologies for the study of molecular processes in biology. With the growing availability of large data sets we anticipate and observe that there is a growing dialogue between these research fields within the genetics research community. Therefore, we are possibly positioned at a point in time where there is a broadening of the tools that will be used in the field of quantitative genetics. As one example of such an approach we will use equation (8) in combination with published QTL results to examine expected response to selection for breeding strategies.

Landscape Metaphor

Sewall Wright introduced the idea of representing the relative performance of populations of genotypes using a landscape metaphor. He focused on evolutionary processes and therefore referred to these as fitness landscapes. From a plant breeding perspective we will generally prefer to refer to these as phenotype or performance landscapes and in some cases adaptation landscapes. In Wright's framework, high-points on the landscape represented regions of genetic space where individuals had high fitness and low-points represented regions of low fitness.

The metaphor of the shape of the genotype-phenotype space as a landscape has been widely used in applications of quantitative genetics to the study of evolutionary processes, but has only been used in a more limited manner to study plant and animal breeding processes. Kauffman applied the concept of fitness landscapes to study features of the genotype-to-phenotype relationship for the *NK* model. Kauffman structured his genetic space by organizing genotypes into genetic neighborhoods based on numbers of alleles shared by the genotypes. Thus, all genotypes are arranged in what can be considered a hypercube and are one step away from all of their possible one-mutant neighbors.

A fitness value is computed for each genotype and the smoothness or ruggedness of the landscape is a function of how the fitness values change with steps between genotypes on the hypercube. We recognize that the landscape metaphor does not work for everybody and may not be appropriate for all situations, but we choose to use it while remaining vigilant of its limitations. Applying the fitness landscape framework of Kauffman, a single genotype is considered a vertex (an intersection point) within the hypercube that defines *N*-dimensional genetic space. Thus, a population of genotypes may be considered as a population of such vertices.

The outcomes of the selection process may be viewed as the creation of a new population of vertices from an old population of vertices within the confines of the *N*-dimensional genetic space. This concept of genetic space may seem theoretical and abstract to the applied breeder. However, this framework can be implemented within a computer simulation environment using equation (3) to investigate any plant breeding process for a range of simple to complex GP models of traits.

SIMPLE TO COMPLEX TRAIT GENETICS

A key motivating principle for organizing and intensively exploring the GP modeling problem space on a complexity continuum is to gain some theoretical insights into the differences in power that can be expected of alternative breeding strategies for a range of situations that simulate simple to complex trait genetics. Using the *E(NK)* model as defined in equation (3), levels of *E*, *N* and *K* are selected and combined with a range of levels of heritability to stratify the complexity continuum in ways that are considered relevant to experimentally determined features of GP models.

The power of different breeding strategies to achieve progress from defined starting points on the performance landscape and thus improve the trait phenotype by searching the performance landscape is compared across the levels of *E*, *N* and *K* and heritability considered. Here we use Kauffman's landscape concept in combination with the *E(NK)* model to examine how the shape of the trait phenotype landscape changes with the genetic architecture of a trait, as determined by changes in the levels of *E*, *N* and *K*. By systematically changing the components of the *E(NK)* model we are attempting to simulate some of the context dependent properties of GP associations for traits as a continuum that ranges from simple to complex.

The simple additive finite locus models are defined by the case where $E=1$ and $K=0$, thus $E(NK) = 1(N{:}0)$ (Figure 1). As

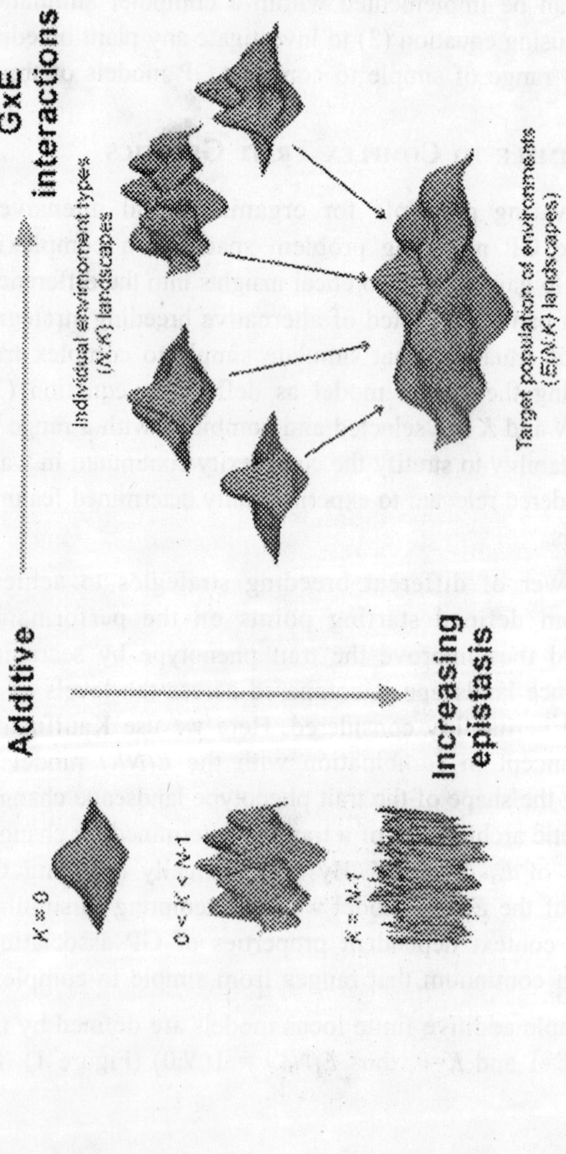

Figure 1: Schematic of performance (adaptation) landscapes for GP models simulated using the E(NK) model.

E and *K* are increased for a given level of *N*, the effects of the alternative alleles for the *N* genes become increasingly context dependent on the genotypes of other genes and on the range of environment-types in the target population of environments. Thus, context dependent effects of genes due to epistasis and gene-by-environment interaction can be simulated.

Building on the landscape metaphor (Figure 1), we observe that as *E* and *K* are increased we move from a single peaked additive landscape for the *E(NK)* = 1(*N*:0) case to a multiple peaked landscape and ultimately a random landscape when $K=N-1$ and $E>1$. Kauffman discusses the shape of a landscape in terms of its ruggedness. We can quantify the complexity of the landscape shape by computing an autocorrelation coefficient between phenotype values for the neighboring genotypes in the genetic space for sequences of random walks across the landscape.

A high value of the autocorrelation is associated with a relatively simple landscape structure and conversely a low value is associated with a complex (rugged) landscape (Figure 2). While we are interested in the shape of the landscape for a GP model we are also interested in how effective different plant breeding strategies are at moving populations of genotypes across features of the different landscapes and in particular their power to move populations to positions of higher performance on any landscape. Thus, we construct a complexity-response plot for any GP model by plotting the landscape autocorrelation coefficient against the response to selection for a breeding strategy.

The response can be measured in a number of ways; here we measure the change in population mean phenotype performance after a number of cycles of phenotypic (mass) selection for an ensemble of *E(NK)* models (Figure 2a). Using the complexity-response plot we observe a quantitative representation of the expectation that as the complexity of the

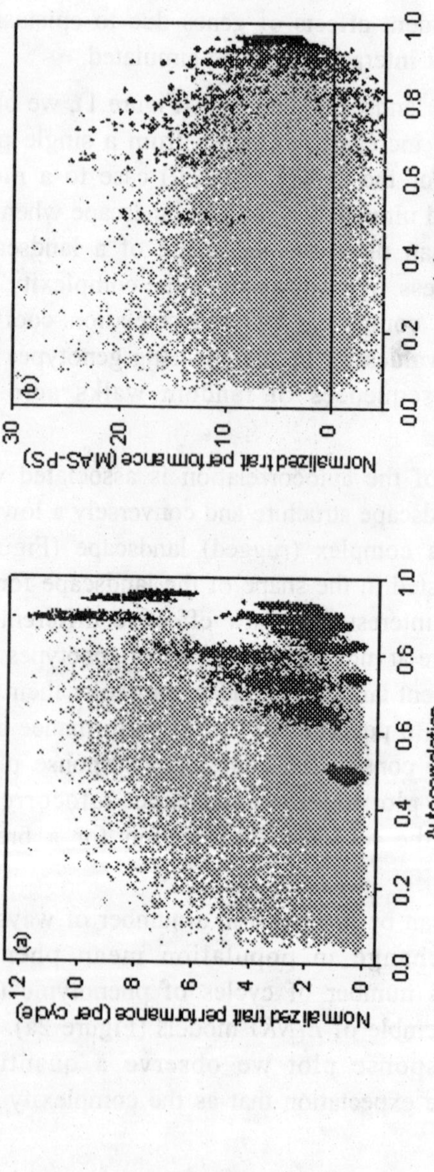

Figure 2: Complexity-response plots for an ensemble of E(NK) genetic models; (a) Response to phenotypic selection (change in population mean per cycle after five cycles), (b) The difference between marker-assisted selection (MAS) and phenotypic selection (PS) at cycle five (b).

genetic architecture of a trait increases the response to selection tends to decrease. Equally we can construct a complexity-response plot that shows the difference in the response to selection between two breeding strategies (Figure 2b). In this example marker-assisted selection (MAS) is compared to phenotypic selection (PS).

While the applications of the *E(NK)* model are defined to simulate putative properties of the ways genes interact in networks to determine trait phenotypes, it can be argued that there is nothing intrinsically biophysical about the genotype-to-phenotype mappings specified by any ensemble based parameterization of the *E(NK)* model. By drawing the effects of the gene combinations from some underlying distribution of effects we have defined genotype-to-phenotype mappings without any need to incorporate specific biological phenomena. Wherever it is feasible we seek to replace the artificial ensemble of *E(NK)* models of GP relationships with experimentally determined and validated GP models for traits.

Given the diversity of the GP problem space such models will take many forms. At present some of the options available include: (1) direct parameterization or replacement of the *E(NK)* model using the results of QTL mapping studies for specific traits, i.e. equation (7); (2) molecular network models; (3) integration of separate trait mapping studies using appropriate ecophysiological models and crop growth and development framework. Here we use equation (8) to superimpose some preliminary published QTL models for traits on the problem space depicted by the complexity-response plot given in Figure 2.

Figure 3 shows the results of a simulation experiment where the complexity-response plot was constructed for two traits using the results reported from QTL mapping studies (Figure 3a). In both examples, the explained genetic component was parameterized by the QTL information reported in the mapping

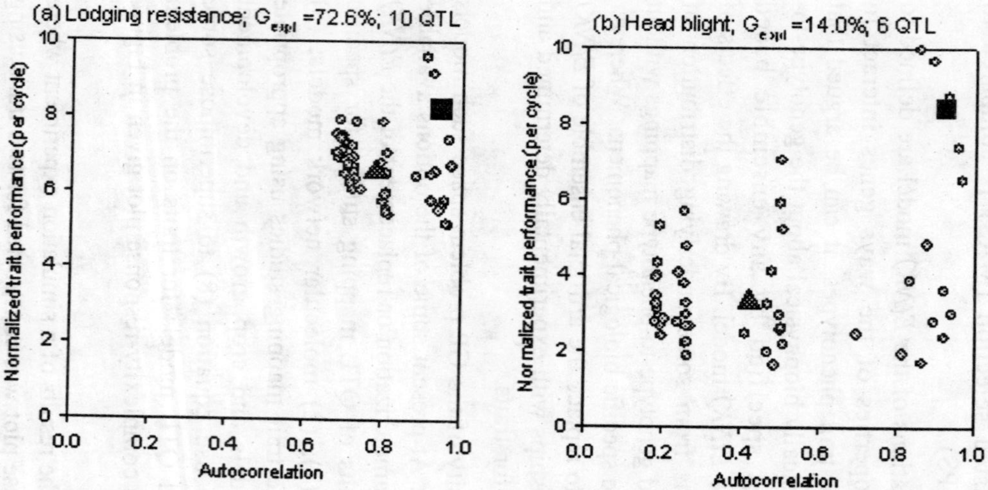

Figure 3: The complexity-response plot for two traits, where the "unexplained" genetic component is defined by a range of E(NK) models. The circles indicate the results from individual E(NK) model parameterizations. The solid squares represent the average results from the E(NK) model parameterizations that contained additive effects only. The solid triangles represent the average results from the E(NK) model parameterizations that contained contain epistasis, gene-by-environment interactions, or a combination of both (Figure 2a).

study, and the unexplained genetic component was parameterized by an ensemble of *E(NK)* model effects following the example given in equation (7). For the traits considered here, the number of QTL in the explained component of the model was 10 additive QTL (Figure 3a; Lodging resistance) and 6 additive QTL (Figure 3b; Head blight).

Based on the results of the published QTL mapping studies, it was assumed that these QTL *explained* 72.6% and 14.0% of the genetic variation for the two traits, respectively, and the remainder of the genetic variation was generated by the *unexplained E(NK)* model component of equation (8). In this experiment, 50 different *E(NK)* model scenarios were considered for the *unexplained* genetic component giving rise to 50 different parameterizations of equation (8) for each trait. The *E(NK)* models considered were a factorial combination of *E*=1, 2, 5, 10; *N*=2, 5, 12, 24, 36; and *K*=0, 1, 2, 5. Thus, the unexplained component of the genetic model ranged from simple (*E*=1, *K*=0; *i.e.* additive effects only) to complex (*E*=1, *K*>0; *i.e.* epistatic effects; *E*>1, *K*=0; *i.e.* gene-by-environment interaction effects; *E*>1, *K*>0; *i.e.* gene-by-environment interaction and epistatic effects).

For each of the 50 parameterizations of trait performance, a complexity-response plot was constructed by considering trait response after five cycles of directional selection and computing the autocorrelation coefficient for the generated performance landscape, as described above. The performance landscape was defined by the combined effects of the *explained* and *unexplained* genetic components of equation (8). Grey circles in Figure 3 represent the complexity-response values for each of the 50 parameterizations of trait performance.

The spread of circles on each of the figure panels (Figure 3) illustrates the potential variation in the outcomes for different genetic models, in terms of the structure of the performance landscape and the expected response to selection. For the lodging resistance trait (Figure 3a), a large percentage of the genetic variation is *explained* by (known) additive QTL effects. Hence, there is a relatively small spread in the positions of the circles and the circles are confined to the upper-right portion of the complexity-response plot, indicating high levels of response and a relatively smooth performance landscape.

In contrast, for the head blight trait, a relatively small percentage of the genetic variation is *explained* by (known) additive QTL effects and hence trait performance is dominated by the "*unexplained*" *E(NK)* model component of the genetic model. This results in a large spread in the positions of the circles (Figure 3b). Furthermore, there is a large difference in the complexity-response values depending on whether the *unexplained* genetic component is defined as simple (*i.e.* additive; solid square) or complex (*i.e.* epistasis, gene-by-environment interaction; solid triangle). Based on the results of this experiment, we would have significantly less confidence in predicting the outcomes from selection for the head blight trait compared to the lodging resistance trait, assuming the results of the mapping studies are representative of the genetics of these two traits.

The approach described above can be applied to the results of any QTL mapping study. Figure 4 shows the results of a simulation experiment where the above approach was applied to 130 different mapping studies reported in the plant breeding literature. The large variation in the outcomes emphasizes that the genetic architecture of traits can be considered a continuum from simple to complex. As illustrated by the clustering of additive models in the upper-right portion of the figure, assumptions about additivity can result in an optimistic view of the potential responses to selection for complex traits.

However, a lower and perhaps more realistic representation of the expected responses to selection for complex traits is observed when different forms of context dependency (i.e. epistasis, gene-by-environment interaction) are introduced into the genetic model (Figure 4). In some cases the assumptions of

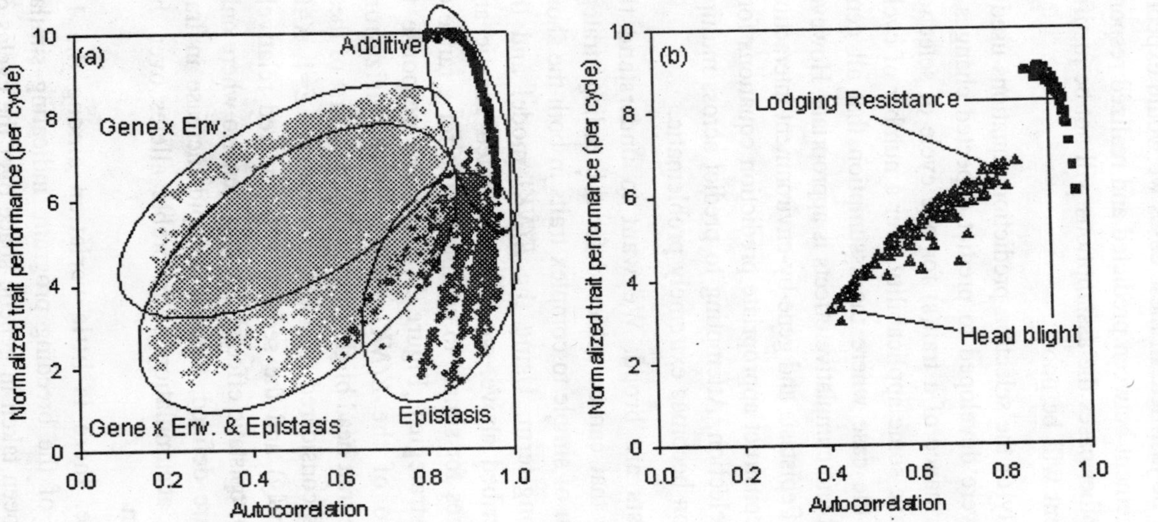

Figure 4: The complexity-response plot based on the results of 130 mapping studies, where the "unexplained" genetic component is defined by a range of E(NK) models (cf. Fig. 2). The results from individual E(NK) parameterizations are shown in (a). The average results from the E(NK) model parameterizations that contain additive effects only (squares) or epistasis, gene-by-environment interactions or a combination of both (triangles) are shown in (b).

additivity or specific forms of epistasis and gene-by-environment interactions may be justified. In these cases we would expect to see a closer agreement between predicted and realized response to selection. In other cases the assumptions will not be justified and the agreement will be poor.

The majority of the selection prediction equations used in plant breeding were developed to predict expected changes in mean trait performance of a trait(s) for one cycle of selection. The equations have some applicability over a number of cycles of selection for the case where the assumption that all genes have independent and cumulative effects is appropriate. However, in the presence of epistasis and gene-by-environment interactions it is difficult to construct appropriate prediction equations for a single cycle of selection. Attempting to predict across multiple cycles of selection becomes extremely problematic.

Our interests are broad. We want to understand the predictive power that can be achieved for a breeding strategy for the continuum of simple to complex traits in both the short-term and the long-term. Using the *E(NK)* model and the components described above, we have simulated genetic and phenotypic changes for simple to complex traits for a range of plant breeding strategies. Figure 5 shows the response to selection of two of the *E(NK)* model parameterizations considered for the trait head blight (Figure 3b). The first of these genetic models is considered relatively simple (i.e. E=1, K=0; additive effects only) and the second is considered relatively complex (i.e. K>0; epistatic effects). For the scenario where only additive effects are defined, there is a rapid increase in trait performance and accumulation of favorable alleles over the cycles of selection.

Furthermore, there is little variation among the 10 independent runs of the breeding program, indicating similar trajectories have been taken in genetic space over the cycles of selection. For the model with epistatic effects, the phenotypic and genotypic response profiles are typical of what we observe

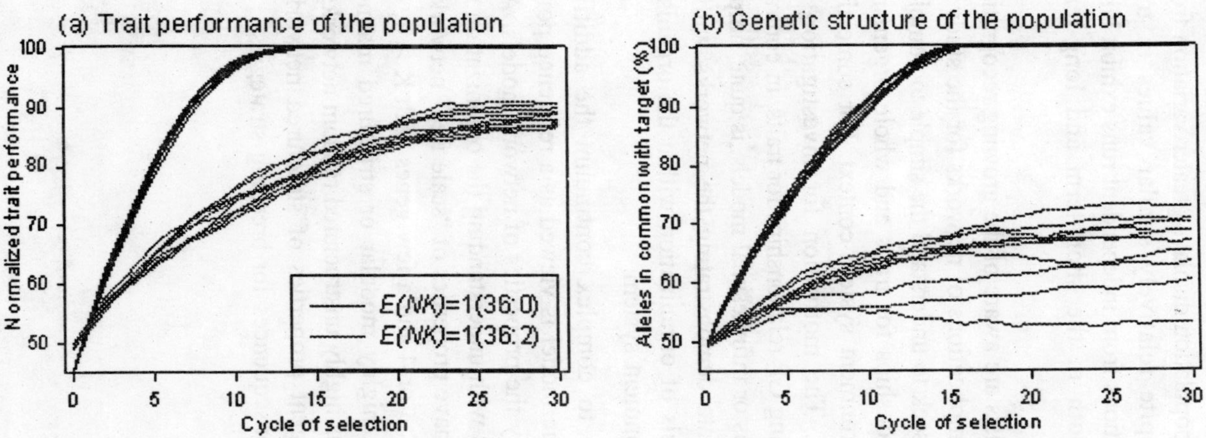

Figure 5: Response to selection for two genetic models. Each line represents an independent run of a breeding program from exactly the same starting reference population of germplasm.

for multiple peaked performance landscapes. Here, the response to selection is much slower than for the additive model and the genetic structure of the population displays greater variation from run to run. Thus, despite relatively similar values in trait performance, the populations from independent runs exhibit large genetic differences, both in the short-term and long-term responses to selection.

Today most scientists are aware of the growing recognition of the importance and implications of networks for the study of biological systems. We seek to understand the simple to complex continuum of GP relationships for traits and whole organisms within a genotype-environment system context. For some this goal seems unrealistic. The motivation for investigation of network models underlying GP relationships for traits, in contrast to the additive finite locus or infinitesimal models, is quite simply the growing body of data demonstrating the networks within the cell and higher levels of organization within the organism and the genotype-environment system.

Along the simple to complex continuum the additive independent locus genetic model is viewed as a reference point from which we can study the properties of network models. An important question deals with understanding the organization of these networks, do they have properties of "scale free" networks, i.e. many genes with $K < 3$ or 4 and a few genes with $K > 4$, or are they organized in a highly modular or structured manner with "canalization" versus highly unstructured random networks? We might expect emergent properties of structured networks could have important consequences for breeding strategies.

CHAPTER 12

ENVIRONMENTAL SAFETY OF GMOs

Before the appearance of GMOs there have been harmful effects from some of the accidental releases of organisms from laboratories. In 1958 tobacco blue mould (Peronospora tabacina) was brought into the UK for a research institute. In that year the mould spread to four other institutes, including one in the Netherlands, and to a commercial tobacco crop in England. In the following year the disease appeared in the tobacco fields of Belgium and the Netherlands, from where it spread quickly across the rest of Europe.

After several years of crop breeding resistance was increased, but it is a powerful example of the risks of accidental release of new organisms. The recent release of some agriculturally important bacteria into the environment, several years after it was first planned, highlights the growing ease with which scientists now regard some types of genetic manipulation. No major problems have arisen, but registration of work and containment levels are useful requirements. Some examples of the current concerns include:

- spontaneous mutations in pure and mixed cultures when growth conditions are changed
- toxins produced in thermophilic systems
- modification of viruses during fermentation

— cloning of toxic genes and the introduction of antibiotic resistance genes into microorganisms not known to naturally acquire them.

The principle problem now in all work is not mechanical, but of laboratory discipline. Safety committee's exist in all major laboratories, but there is still room for laxity. Medical surveillance of laboratory workers should be more common in case there may be long term affects of exposure.

There are no known health hazards that are specific to genetic manipulation. However, some of the organisms used, especially microorganisms, are capable of infecting humans so can present health hazards. There could be several ways in which genetic manipulation could result in hazards to humans. A GMO which facilitates delivery of a biologically active gene product to a specific tissue could occur. Autoimmune disease resulting from production of antibodies to a GMO or gene product,. Or enhanced immune response to proteins that are fusion products of the proteins that one has become immune to.

In Britain, the Health and Safety Executive is trying to implement a scheme for surveillance as recommended by the U.K. Advisory Committee on Genetic Manipulation. It advises that people working with certain categories of GMO should carry cards, and also have cards attached to their medical records, that would allow contact with a supervisory medical officer should they get sick, as one way of monitoring any effects. It also recommends the storage of serum samples before conducting such work, to allow comparison at a later date in the event of illness. There is still controversy over the use of genetic engineering in industry in some countries. A West German court ruling in November 1989 forbid the use of these techniques in industry until there is a better legal basis for regulation. This forced Hoechst AG to stop the construction of a factory for two years, until recently when a law was introduced.

The application will be considered now that Germany has a gene technology law. The West German law also fits with an

EEC directive on genetic engineering that provides minimum protection in EEC countries without regulations,taking effect in 1991. The German's are particularly sensitive over the use of genetics. Industrial size means that any problems could be on a larger scale, and the large number of GMOs increases the chance of release. The sterilisation of waste products is costly, but very important.

The points to be considered in the safety assessment include characteristics of parental and modified organisms, health considerations such as pathogenicity, and environmental considerations. There should be methods available for decontamination of areas in the event of accidental releases. There are three classes of operation, depending on the organism. Many regulations are written principally in terms of microorganisms, but they need to include plant and animal cell culture also. They must also include the laboratory growth of transgenic plants and animals, and define what is contained or not. The disposal of such organisms should be performed in a similar way to microorganisms. Larger organisms do have the advantage that they are not spread as easily as microorganisms, but pollen from plants or animal insects could be potential vectors.

Biotechnology is advancing into areas that depend on the introduction of genetically modified organisms into the environment. The deliberate environmental introduction of any new organism, including GMOs, should be only undertaken within a framework that maintains appropriate safeguards for the protection of the environment and human health. Natural habitats already contain their own indigenous populations of organisms, organised in a delicate web of nature, which needs to be maintained.

The first experiments are being conducted in as closed environmental situations as possible. Past experience with GMOs is principally in controlled situations, and we must be careful about shifting to new environments. The initial experiments on plants and animals are in enclosed research areas, but when they have proved safe and are economically useful, then they will need to be grown in large quantities. It may be feasible to use enclosed farms for some animals, especially such as fish or chicken farms, or even pigs or cattle. To be of a major practical use to worldwide agriculture, any GMO must be released into the environment.

Only small scale agriculture can be conducted in closed environmental systems, though some important products used today are produced in that way, such as eggs from battery farming of chickens. There have been many field trials since 1984 when the Canadians field tested a transgenic plant. There is widespread public concern about the free release of recombinant organisms into the environment, and questions about how many organisms constitute a significant environmental release. The degree of care required depends on the potential, or known, risk to the ecological balance and humans.

For plants, to be economically significant they will need to be grown over large areas, so that will mean free release on a large scale. If the food is to be grown where it is especially needed in the developing countries, both the economic situation and the social structure of small-scale peasant farmers, will mean completely free release is required. Serious problems have arisen from the unexpected results of the movement of weeds and insect pests into new environments. Some were deliberately introduced as pests into new environments, some introduced to solve one problem, but caused another. Some previous releases of organisms into new environments have proved to be beneficial, or at least harmless, but others have been deleterious.

The question of environmental release of GMOs is applicable to the release of bacteria, plants, animals and humans. The possibility of a novel organism or virus being widely dispersed, and/or disruption of the ecosystem, is the main fear, and in view of the dramatic consequences possible, it is very serious. It has been considered serious enough for several countries to ban any introduction for testing already. It is a rapidly increasing problem as the technology is very cheap and becoming well known, so that many people can try different modifications. In view of the important benefits that new organisms can offer innovation should not be discouraged.

DEGRADE ENVIRONMENTAL POLLUTANTS

Bacteria can be made to degrade environmental pollutants. There are a range of bacteria available that can metabolise, or literally live off substances while removing them from the environment, including many important pollutants such as polychlorinated biphenyls, dioxin, herbicides, pesticides and oil spills. The first patents obtained for a microorganism involved the engineering of Pseudomonas strains to degrade chemicals found in oil spills. The genes enabling degradation of environmental pollutants can be introduced into different bacteria. For instance the ability to degrade toluene has been transferred into bacteria that can live a zero degrees Celsius. Many of the bacteria used to supply the genes are actually discovered at hazardous waste sites, after naturally evolving to cope with the chemicals present. This pollution problem is quite serious, so this could be a major use.

A bacteria that can reduce the concentration of trichloroethylene by a thousand times has recently been made. Trichloroethylene is one of the most significant environmental pollutants, and a suspected carcinogen . The use of bacteria to reduce the level of it in drinking water could be very important. There are other bacteria that can be used to extract and concentrate heavy metal containments from places such as land fills, mine tailings, or low grade mineral ores. There are organisms that can extract cobalt, and research to produce mercury concentrating bacteria. One study found that heterogeneous microbial populations may be better than single species systems. The organic matter needed to act as a metal-chelating biomass can be reused, with autodigestion of excess biomass.

Acclimation to high levels of one metal, e.g. cadmium, did not confer acclimation to another metal, e.g. copper. Genetic engineering will be needed to insert multiple genes to allow recovery of the mixture of toxicants at many sites. These metals are also valuable to recover, so the work has dual benefits. High value products, like precious metals, provided some incentive to the process, and there are commercial products available. There are also organic pollution problems which may be treated by using bacteria, one of the most important being wastewater treatment schemes. It is also possible to use fungi for some of these applications. The fungi are attached to woodchips which can be mixed with the polluted soil, to metabolise pollutants.

In Britain there have been several field trials of GMOs. One series has involved baculovirus insecticides, which are viruses that only infect and kill a few species of insect. They have no effect on other types of insect or other species, and do not pollute the environment. Naturally-occurring baculoviruses have been used during this century, and more than a dozen have been employed commercially.

The bacteria Bacillus thuringiensis has been used for twenty years as a pest control, and is standard practice for the control of mosquitoes. The spores can be dispersed in a water mixture and distributed. There are over 400 registered formulations of B. thuringiensis in the USA. There are a number of insecticidal proteins produced, the two most important are the beta-exotoxin and the delta-endotoxin. The beta-exotoxin is effective against a wider range of insects, and has a greater soil persistence, but it may be mutagenic and it is not permitted to be used in the USA. In Europe it is used in the USSR only. The strain used to produce delta-endotoxin is usually B. thuringiensis var. kurstaki.

The protoxins are laid down as crystals in the spores, and only after ingestion, when the protoxin is hydrolysed in the insect gut does the active protein form. It damages the membranes of the cells lining the larval gut; the insects stop eating within minutes, and are paralysed, dying within two days. There have been several insecticides released using B. thuringiensis insecticidal proteins. Many of the preparations contain live bacteria. Mycogen has a product called M-One which contains a variety of the bacteria. There is only one reported case of resistance to this protein, and it was in a limited situation. If resistance does appear then other strains can be substituted. However, B.thuringiensis does not persist well in the environment, and the spores are sensitive to UV light, so the endotoxin gene has been expressed in other bacteria, which are being field tested. This would reduce the number of applications needed. The B. thuringiensis insecticidal protein genes have been cloned into a variety of bacteria.

One test case was the application by Monsanto to field-test a soil bacteria (Pseudomonas fluorescans) which has been engineered to produce this endotoxin. One species of root-colonising Pseudomonas with the toxin gene can be used to treat seeds to make them resistant to cutworm. The toxin gene has been put into an endophytic bacterium, Clavibacter xyli subspecies cynodontis, that colonizes the xylem of plants to make the plants resistant to Lepidoptera. A field trial is being conducted. It has also been introduced into blue-green algae which might serve as food for mosquito larvae in ponds and streams. This is in addition to the insertions directly into the genome of higher plants.

One solution to the release of live genetically engineered bacteria is to use dead bacteria. In September 1987 the U.S. company Mycogen received U.S. patents for the invention of a process that kills bacteria while preserving their cell wall as a gelatin-like capsule which remains intact until the insect pests eat them, only then releasing the contents such as a pesticide. Almost any product could be put in the capsules. Several genetically engineered toxins are being field tested by Mycogen and Japanese companies. This invention allowed Mycogen to be the first U.S. company to receive permission from the EPA to field test genetically engineered bacterial pesticides, in 1985. These are alternatives to chemical pesticides which are very damaging to the environment.

TRANSGENIC PLANT FIELD TRIALS

Initially trials of genetically modified plants can occur in closed greenhouses, however many plants respond differently to conditions in a greenhouse and in the field. It is difficult in the greenhouse to simulate natural diurnal and stress conditions, and not all the plants can be grown there. Only when growing under natural conditions is it possible to determine if the genetic trait is expressed, effective, and does not have detrimental effects. Agricultural plants have been the economically most important group of exotic introduced organisms in the past. There have been a few disruptions of natural ecosystems by introduced crops.

One example is the spreading of bamboo from cultivation to cover large areas of the mountain sides of Caribbean Islands. There have been many field trials of plants containing the Bacillus thuringiensis insecticidal protein. They have shown excellent insect control, suffering no damage next to parent plants that have been totally defoliated. Early market opportunities for caterpillar resistance are leafy vegetable crops, cotton and corn. Crop targets for beetle resistance are potato and cotton.

Transgenic tomatoes carrying the tobacco mosaic virus coat protein gene have been shown to be highly resistant to virus infection, with no yield loss. Transgenic potatoes with multiple viral resistance have been tested. Many crops have been tested with resistance to viruses, another example is cantaloupe and squash with resistance to cucumber mosaic virus, papaya ringspot virus, watermelon mosaic virus and zucchini yellow mosaic virus. When we consider that the first transgenic plant was made in 1983, and there were field tests with genetically modified tomatoes and tobacco well underway in 1987, we realise how rapid progress has been. Tobacco is very easy to modify, that is why it is often use, rather than for the interests of tobacco companies. The progress is exponentially increasing.

Genetically-modified crops were tested in Europe more than anywhere else during 1989. The countries that have recorded trials of transgenic plants include Australia, Belgium, Canada, Finland, France, Ireland, Israel, Italy, Netherlands, New Zealand, Spain, Sweden, UK and USA. Plant Genetic Systems alone conducted 30 field tests in seven European countries. Their programme in 1988 involved 13 field trials, more than the total number in the USA that year. The reason for their success includes good cooperation with government agencies and independent research institutes. They have also tested crops in the USA. The Belgium-based company involves a research staff of around 100, but conducts field trials only in joint ventures.

There have been a number of field trials involving plants resistant to the Basta herbicide. There have been field trials involving genetically modified alfalfa, Brassica napus, Lycopersicum aesculentum, Oilseed rape, poplar, potato, sugarbeet and tobacco. There has been a variety of other herbicide tolerant genes tested in different plants. There have also been trials with inserted Bacillus thuringiensis insecticidal protein genes, in potato, tobacco, and tomato. There are already many types of agriculturally important plants that have been grown with genetic modifications in field trials. The range of species is growing, and is limited principally by the ability to regenerate entire plants from genetically transformed cells.

The list in mid 1990 included alfalfa, apple, Arabidopsis, asparagus, bananas, cabbage, cantaloupe, carrot, cauliflower, celery, corn, cotton, cucumber, Douglas fir, flax, horseradish, lettuce, lotus, Medicago varia, Morning Glory, Orchard grass, peas, pears, pepinos, petunias, pinetrees, poplar, potato, rape, rice, rye, soybean, squash, sugarbeet, sunflower, tobacco, tomato, trefoil, Vigna aconitifolia, walnut, white clover.

The only examples of intentional release of GMOs into the environment have been recent and are few in number. Since most GMOs only differ from the parent strain in one or a few genes, they will often behave in a similar way. It is relevant therefore to examine existing cases of the release of organisms into the environment. Many GMOs will probably be less fit than the parent organism. Part of the reason is that they have extra DNA to carry, which will slow the reproductive rate - the excess baggage hypothesis. However, this is not necessarily so, and even if it was it may take many generations to pass before the introduced organism disappears due to decreased fitness if the turnover rate of populations is slow.

In the first field test of an engineered soil bacterium, Pseudomonas fluorescans, with a chromosomally inserted lacZY gene marker, the population sizes of the introduced parent and transgenic organisms were identical for 30 weeks. The Pseudomonas bacteria was applied to wheat plants, and the furthermost migration of the bacteria was 18cm. No transfer was observed to drainage areas. There have even been cases where competitiveness has been enhanced with the possession of foreign DNA.

An example where the recombinant strain has shown enhanced survival in the environment, was the survival of Escherichia coli with a transposon ColE1::Tn5, in a farm environment. A variety of responses are also seen in plants. Some support for the idea of the intrinsic weakness of artificially bred plant varieties has come from experience with modern crops, which are often incapable of survival without human intervention. However, some crops such as potatoes, can become a weed in following crops.

Natural selection will act on all organisms, including GMOs. Genetic novelty provides the raw material on which natural selection acts. Selection after the release of the GMO will tend to increase the fitness of the GMO, by reducing costs associated with novel traits. If increases in fitness occur they will probably increase the population growth rate and biological competitiveness, as well as other ecological effects. These need to be considered in risk assessment. Myxoma virus released in the UK and Australia in the 1950's to control rabbit populations evolved to become less virulent. Likewise, with GMOs natural selection will act only in a way to increase the fitness of the GMO, taking no account of human intentions.

Released microorganisms are highly likely to enter freshwater or marine environments via agricultural run-off, or faecal contamination. There are various environmental factors that affect persistence, including moisture, pH, temperature, nutrient level, sunlight and other organisms. The case of Rhizobium inoculants has provided some data. It has been found that introduced strains may not compete successfully for nodule formation with indigenous populations of Rhizobium. When indigenous Rhizobium populations in soil are high it may be necessary to use a thousand-fold number of inoculants.

Very little is known about the biochemical determinants of competitiveness. If there is no indigenous population, then the inoculants can establish as a complete population. In some cases the GMOs may be intended to persist at a particular level in the environment. If the GMO is going to enter a new ecological niche, it may be safe. Those organisms used for biocontrol may need to remain in the environment at a low level in the absence of a pest outbreak. However some GMOs may be required to die out after use, like microorganisms that breakdown particular toxic wastes for emergency use. There will be a minimum inoculant population required if persistence is desired.

This is also a reason why the deliberate introduction of GMOs to the environment is different to the accidental laboratory introductions that have frequently occured during the course of recombinant DNA experiments in scientific research. It has been argued that given that bacteria can rapidly grow in the environment, even if we introduced a few accidentally natural selection would determine whether they survive or not. However, there is a minimum inoculation population that is required in many instances, so we can not say it is always going to be the same as the case of accidental escape of a few organisms from a laboratory. The types of organism may have very different ecological niches, also.

GMOs should be designed for safety, in addition to their function. This may mean that they include some biological containment attributes to reduce their longevity in the ecological niche to which they are released or which alter their ability to transmit genetic material to other organisms encountered in the habitat. At the same time the GMO must persist long enough to perform its task for which several means have been suggested. Mutations can be introduced, such as those to abolish the ability to synthesise cyclic AMP and its receptor protein, as strains of enteric bacteria with these mutations have a 50% longer generation time.

A gene such as hok which encodes a 52 amino acid polypeptide that mediates a lethal collapse of the transmembrane potential, that kills a certain proportion of each generation or which can be linked to a chemical inducer, has been tested. A plasmid vector that causes the destruction of any recipient cell could also be made. There has been successful use of self-destructive baculoviruses as mentioned earlier. Detection and subsequent elimination of organisms may be feasible if they are large, such as large animals or plants, however, insects, microorganisms or viruses may be difficult to exterminate after introduction.

GENES TRANSFER

Transfer of engineered genes from the GMO to other organisms may occur through hybridisation in higher organisms, or through conjugation, transduction or transformation in microorganisms. If lateral transfer occurs, an engineered gene may persist in the natural environment after the GMO is no longer present. An important unknown question is how often lateral gene transfer occurs. It occurs at a higher rate in the presence of selection pressure in the environment. Among microorganisms the scientific evidence suggests that it is neither so rare that we can ignore its occurrence, nor so common that we can assume that barriers crossed by modern biotechnology are comparable to those constantly crossed in nature.

Current evolutionary thought does not consider that organisms are perfectly adapted to their environment, as there are important restraints for this to occur: namely, the presence of suitable genetic variants, which is what may be introduced in GMOs. Techniques are being developed to reduce the potential for gene transfer, but there is little known outside of experience with a few laboratory species. There are four essentials of enduring lateral gene transfer, and each component is improbable:

— the origin of an unusual genotype (to the recipient)
— its transmission to an extraspecific cell
— a resulting advantage to the recipient cell
— multiplication of the recombinant to numbers that render it safe from random loss.

This scheme has often occured over the long period of evolution. Genetic engineering does increase the chances of gene transfer occuring. If we are generating novel genetic combinations, experimental observation of organism behaviour is important. One possibility is that bacterial viruses could acquire a capacity to infect higher organisms, which may upset the extremely intricate ecological balance. However, some bacteria commonly pick up human genes naturally. The genetic barriers thought to exist between species are in fact often broken, especially by viruses.

The argument against genetic manipulation when we cross species barriers is weak, however, as it has been found that these barriers are broken in nature. Nature has set barriers to horizontal gene transfer in eucaryotes, but trans-species gene exchange by a process called conjugation is common among procaryotes. Another piece of evidence is that plant breeders have used comparatively crude techniques for decades to transfer genes en masse from wild species into crop plants, without adverse consequences. Traditional breeding may move the desired traits together with hundreds of other genes into a new variety. There have not been any cases of detrimental gene transfer from released organisms in the past, so there is unlikely to be problems with single, precisely engineered gene alterations, unless it is an inherent property of the gene vector to move to other organisms under field conditions. If the trait that is being transferred is already in the environment then we will have less concern regarding its transfer, especially if the particular engineered gene is present in the system that the trial is conducted in.

The range of organisms affected by a virus may be altered in less direct ways. A laboratory study of a pathogenic plant virus showed that by altering a single gene, the range of insects that could carry it was changed. This would enable the virus to come into contact with previously unaffected plant species. There needs to be experiments to determine the distance that pollen spreads, and how it hybridises. Plants often show little barriers to hybridisation, so there is always a potential for exotic pollination. Hybridisation has often been observed in sorghum and wild relatives, but there is still no evidence that traits such as resistance to insect pests have spread from traditional crops to wild relatives.

There has been little research conducted on cross-fertilisation of GMOs There has been a study performed at DSIR Crop Research, New Zealand. Transgenic potatoes resistant to chlorsulfuron were grown with a border of wild type potato plants to measure pollen dispersal. The frequency of transgenic seedlings among the progeny of wild-type potatoes growing within the trial was about 1%, but only 5 in 10,000 of the progeny from wild-type potatoes planted up to 4.5m from the trial, were transgenic. There were no transgenic progeny recovered from wild-type potatoes growing 4.5 to 10m from the trial.

Most of the traits bred into domesticated crops by traditional breeding, such as rapid seed germination, would be detrimental to weeds. Many of the new traits will be beneficial in weeds also, such as pest resistance. Even if there is a very limited opportunity for cross-breeding, with large numbers, it is virtually inevitable that gene transfer will occur. Crops often grow in the same region as wild relatives.

Although it may not be considered by many protest groups, the most likely potential ecological effect of the release of GMOs will be to protect the environment from many harmful chemical pollutants. Overall these new genetic technologies promise much to aid world agricultural techniques. If plants were made to use fertiliser more efficiently it would mean less fertiliser would run off into rivers causing pollution, and if they were made disease resistant then less problems would arise from the poisoning of the environment by herbicides. They are cheaper and should help to solve the pollution problems caused by the current fertilisers, herbicides and insecticides. In some cases where the ecosystem is changing due to climatic conditions, GMOs are being developed to maintain a species presence in the area, especially with the modification of trees, which may be required by many dependent insect and bird species. The inherent nature of any inter-species interchange of the genes, the major concern is safety of the GMOs and their genes on a case-by-case level to the ecosystem. Understanding ecological interactions is crucial to the planned introduction of GMOs. Most GMOs will pose minimal ecological risk.

New varieties have been introduced in the past, just as the GMOs will be. If successful the crop will be grown over a large area. Many of the traditionally bred genetic variants are chosen because they have disease or pest resistance, so they have similarities to the GMOs. If the GMO release had an impact on the natural community, the consequence would probably be a transitory disturbance of the community structure. There are, however, some examples in nature where the acquisition of a single gene can cause ecological problems, such as antibiotic resistance genes that have been acquired by bacteria, in the gastrointestinal tracts of animals. From past mistakes we should learn the need to be cautious when applying new technology and introducing new organisms. About 10% of the eExotic species introduced into Britain have become established in the wild, and about 1% have become pests.

The distinction between a crop plant and a weed is very narrow. Of the 18 worst weeds in the world, only one is not used somewhere for an agricultural purpose. There are weed oats, barleys, rices, maizes, carrots etc. A similar situation exists with domestic versus wild animals that are pests. The alternative which is probably more acceptable is to put into the environment only genes that we can tolerate spreading to other species. For instance if the gene for resistance to one herbicide is spread to weeds, then another could be used. One could even imagine the possible industrial sabotage of companies by the introduction of weeds that are resistant to their competitor's herbicides.

There are several methods by which weeds could arise from the introduction of GMOs. One is through hybridisation with local weeds, which was mentioned previously, and which could be the most serious. There is also the low possibility that the GMO could evolve into a weed itself . This sort of risk is probably worth the large economic savings and increased crop production that is possible using herbicide resistant plants. Bacteria can grow very rapidly, so there could be more rapid spreading of new genes among bacteria, such as in the many species found in contact with ground water.

However, with alterations such as the removal of ice-nucleation proteins, serious reasons for concern have not been found so far from the recent results of preliminary field trials. New pests could arise, as for instance in the escape of a salt-tolerant rice cultivar from cultivated fields to estuaries. It is possible for both native or exotic species with new traits to become pests. Sometimes the change in environment means that plants may become major weed problems. For example kudzu in America, where the climate is different, and it has a small competitive advantage, has over several generations spread to become a weed. Release of a genetically modified cultivar in regions which that cultivar has many wild relatives could see hybridisation between the cultivar and wild relatives.

In Africa, sorghum cereal crops have hybridised with weedy relatives to produce a serious pest known as "shattercane" which can mimic the crop. Enhancement of the effects of existing pests could occur if they hybridised with GMOs. Weeds could acquire herbicide resistance, or insect resistance, or other advantageous traits. Nontarget species might be harmed, such as insects other than the insect pests. The genes in farm animals might be transmitted via faeces. There has been a case with a recently applied antiparasitic drug, ivermectin, which is currently applied widely to cattle, horses, sheep and pigs in many countries. It has been found to have an insecticidal effect on the larvae of a fly which normally helps break down the dung. Most of the applied drug is quickly passed in the faeces, and the high concentration of the drug kills the insect.

As a consequence, dung was not decomposed quickly enough, and it killed the pasture where it accumulated. One frequently expressed concern is the potential for GMOs to displace resident species in the receiving community, particularly microbial species performing key functional roles such as nitrogen fixation or lignin decomposition. Because redundancy of function appears to be common in microbial communities, in many cases there would be little concern over microbial species replacement caused by an introduced GMO. This effect would have the result of altering the diversity of species in the environment.

The possibility of organisms developing resistance to GMOs is very real, and can be expected. The probability of it occuring will vary, as will the significance. We should plan strategies based on the assumption that resistance will develop, at some stage, in some trial or commercial application, and have a followup strategy.

A study in Southern Italy has shown how there have been some mutations induced by the vaccine to enable Hepatitis B virus to infect children that had been vaccinated. A total of 44 people, including infants of carrier mothers, became HBsAg positive, that is positive to hepititis B surface antigen, despite passive and active immunisation against it. In one infant this led to serious disease. The virus can mutate so that it loses the determinant that the vaccine binds to, so the remainder of the virus can infect the patient. There is genetic variation in the herbicide resistance that has been found to occur in nature.

Multiple herbicide resistance has arisen in the wheat weed Alopecurus myosuroides after selection due to treatment with a single compound. The mechanism of resistance is not understood which should make us very cautious in the introduction of large amounts of herbicide-tolerant plants. In the laboratory it is possible to isolate mutant varieties of corn or tobacco or other plants resistant to sulfonylureas and imidazolinones, often the result of a single dominant mutation. Careful management is required, and broad farming strategies should be examined.

Organisms that have new phenotypes can provide selective pressure on other organisms. The evolution of resistance to the Bacillus thuringiensis insecticidal protein following the introduction of these crops has been modelled. The model suggested that the presence of the insecticidal protein will be a powerful selection pressure, which could result in acquisition of resistance. There is also one situation where insect pests have evolved resistance to the insecticidal protein, when seeds stored in a silo were dusted with the spores.

Resistance to the endotoxin has never been observed in nature, despite its continued use for many years. The spores are sensitive to UV light and degrade after several days, releasing the protoxin into the environment where it is broken down. However, if it is continually present, as it would be in transgenic plants, insects have an increased chance of becoming resistant. This has been achieved in laboratory selection experiments using Indian meal moth, where the lethal dose can be increased by 250 times compared to parent strains. Lower levels of resistance have been obtained in almond moth and tobacco budworm. The mechanism of resistance has been found to be due to alteration in toxin-membrane binding. If this did occur it would reduce the efficacy of new resistant crops, and the efficacy of current uses of the toxin, and it could also change the role that the insecticidal protein plays in the natural ecosystem.

Modelling studies have shown that the time needed to develop resistance to the toxin could be significantly extended if some unprotected crops were planted in the mixture, as this increase in genetic variation is an important factor. This is an example of efforts directed at using the toxin genes in a way so as to reduce the rate that pests evolve resistance. This would limit the selection pressure on insects for the development of resistance. The probability of pests becoming resistant could be more substantially reduced if several genes were simultaneously used in the same plant. Combinations of insecticidal protein and protease inhibitors could be used. In a similar way, several methods of viral protection could be used to avoid viruses overcoming plant resistance.

The combinations could be changed each year to ensure the pests were not able to overcome such methods, and the techniques of genetic engineering will allow this rapid breeding. During the last few years, 60% less insecticides have been used in US cotton fields in attempts to manage the selection of pests resistant to chemicals. Resistance to one variety of toxin can be countered by application of a different variety of B. thuringiensis with a variant toxin.

CHAPTER 13

CRITICAL ROLE OF PLANT BIOTECHNOLOGY

Within the last 100 years the world has seen the rise of genetics as a scientific discipline, the finding of DNA as the hereditary material, the elucidation of the double helix structure of the DNA molecule, the cracking of the genetic code, the ability to isolate genes, and the application of DNA recombinant techniques. Methods of crop improvement have also changed dramatically throughout this century. Mass and pure line selection in landraces, consisting of genotype mixtures, were the popular breeding techniques until the 1930s for most crops. In the 1930s maize breeders started the commercial development of double cross hybrids that was followed by the extensive utilization of single crop hybrids since the 1960s.

Pedigree-, bulk-, backcrossand other selection methods were also developed especially for self-pollinating crop species. Such scientific advances in plant breeding led to the so-called 'Green Revolution', one of the greatest achievements to feed the world in the years of the Cold War. Owing to this agricultural betterment, cereal production, which accounts for more than 50% of the total energy intake of the world's poor, kept in pace with the high average population growth rate of 1.8% since 1950.

Today, 370 kg of cereals per person are harvested as compared to only 275 kg in the 1950s; i.e., in excess of 33%

per capita gain. Similar progress in other food crops resulted in 20% per capita gains since the early 1960s, according to FAO. There are 150 million fewer hungry people in the world today than 40 years ago, though there are twice as many human beings. Despite this splendid progress in crop productivity, even greater progress must be made in order to feed an additional two billion people by the early part of the 21st century. Careful choice of such biotechniques as well as a realistic assessment of their potential in crop improvement are needed to avoid not only the criticism of the anti-science lobbyists but also the permanent distrust of pragmatic traditional breeders.

For example, a World Bank panel recently released for discussion a well based report concerning bioengineering of crops. The panel members recommend "to give priority to all aspects of increasing agricultural productivity in the developing world while encouraging the necessary transition to sustainable methods". Indeed, plant biotechnology has been regarded as a priority area for technology transfer, because genetically modified food, feed, and fibre are of vital concern to the developing world. Therefore, the rich industrialized world should share their biotechniques and avoid policies that do not allow the progress of agriculture in poor, nonindustrialized parts of the world, where this economic activity still provides 60 to 80% employment and 50% of national income.

Such support will assist the developing world towards food self-reliance, which will be very important to avoid hunger and keep peace in many regions of the tropics, where the agricultural sector remains the most important basis for economic growth. Furthermore, a wealthy society provides high living standards to its citizens. Tissue culture was developed in the 1950s and became popular in the 1960s. Today, micropropagation and in vitro conservation are standard techniques in most important crops, especially those with vegetative propagation. At the beginning of the 1980s genetic engineering of plants remained a promise of the future, although gene transfer had already been achieved earlier in a bacterium.

The first transgenic plant, a tobacco accession resistant to an antibiotic, was reported in 1983. Transgenic crops with herbicide, virus or insect resistance, delayed fruit ripening, male sterility, and new chemical composition have been released to the market in this decade. Argentina is the leading developing country with an excess of 4 million ha of transgenic herbicideresistant soybean. There are 4.4 million ha of transgenic corn, 5 million ha of transgenic soybean, and 1.6 million ha of transgenic canola grown only in North America. It has been calculated that in 1998 US farmers are growing over 50% of their cotton fields with transgenic seeds, the largest percentage for any crop ever.

Allozymes were available as the first biochemical genetic markers in the 1960s. Population geneticists took advantage of such marker system for their early research. In the 1970s, restriction fragment length polymorphisms (RFLP) and Southern blotting were added to the tool box of the geneticists. Taq polymerase was found in the 1980s, and the polymerase chain reaction (PCR) developed shortly afterwards. Since then, marker-aided analysis based on PCR have become routine in plant genetic research and marker systems have shown their potential in plant breeding.

Furthermore, new single nucleotide polymorphic markers based on high density DNA arrays, a technique known as 'gene chips', have recently been developed. With 'gene chips', DNA belonging to thousand of genes can be arranged in small matrices and probed with labeled cDNA from a tissue of choice. DNA chip technology uses microscopic arrays of molecules immobilized on solid surfaces for biochemical analysis.

An electronic device connected to a computer may read this information, which will facilitate marker-assisted selection in crop breeding. In summary, since Mendel's work on peas, there have been five eras in genetic marker evolution: morphology and cytology in early genetics, protein and allozyme electrophoresis in the pre-recombinant DNA time , RFLP and minisatellites in the pre-PCR age, random amplified polymorphic DNA, microsatellites, expressed sequence tags, sequence tagged sites, and amplified fragment length polymorphism in the oligoscene period, and complete DNA sequences with known or unknown function as well as complete protein catalogs in the current computer robotic cyber genetics generation. The driving force for such a development has been the scientific interest of human beings to understand and manipulate the inheritance of their own characters.

The advances in plant transgenics and genomics have not been isolated from society. Some of these achievements have been acclaimed by end-users whereas other accomplishments, e.g. release of genetically modified organisms (GMO), are being attacked, not only in words but also in deeds, by political activists. Some of these educated middle-class campaigners are expressing in this way their rampant 'eco-paranoia', while others hide their real agenda to manipulate the fashionable ecological movement. This controversy has attracted the attention of non-scientific partizans to each side.

There have been negative comments about transgenic plants by a crown prince and contrasting positive comments by a former president, both of whom may not have the required technical knowledge to assess the potential of biotechnology for crop improvement. Irrespective of this ideological dispute and ensuing democratic disagreements, biotechnology products will be accepted by people who support scientific-based progress, in a similar way that new cultivars or innovative crop husbandry techniques have previously become integral parts of farming systems elsewhere. However, without end-user's consent, the impact of a new technology in the society will be small or nil.

Scientific honesty seems to the best policy to convince people about the advantages of biotechnology for crop improvement. Scientists, farmers, consumers, and policy-makers should objectively assess the potential hazards of crop biotechnology in farming and food systems regarding the current situation and the likelihood that such hazards may occur. For example scientists should explain to the people that gene recombination already occurs in nature. However, the ecological success of viable recombinants after gene reassortment is unpredictable owing to the high fitness of current isolates. For this reason, more scientific research will be needed to identify unpredictable risks and the chances of their occurrence.

The need for profit, as in any other business, has attracted the interest of the private sector to defend their investments in crop biotechnology with patents, intellectual property rights, and new protection methods, e.g. 'terminator' technology that inhibits germination of self-pollinated seeds. This technology protection system prevents farmers from saving seeds from their harvest for further utilization as next season planting propagules. Three genes, each with a specific promoter, are inserted into the 'terminator' plant.

One of the genes produces a recombinase that removes a spacer between the gene producing, for example, a ribosomal inhibitor protein and its promoter such as late embryonic abundance, which only becomes active during the late stages of embryo development. This spacer with specific recognition sites blocks the gene from being activated. Another gene produces a repressor that keeps off the recombinase gene until an outside stimulus is applied to the 'terminator' plant, e.g. a chemical such as the tetracycline, or temperature and osmotic shocks. The United States Department of Agriculture (USDA) and a cotton seed enterprise jointly acquired a patent for this concept.

Two months after this patent was announced, one of the leading agro-chemical transnationals bought the cotton seed company, although one of its officers said that it may take many years before this 'terminator gene' idea becomes a proven technology in the seed industry. Strategic alliances, joint ventures, research partnerships, new investments, company mergers, cross-ownerships, and take-overs in the seed and agro-chemical business have also been in the news in recent months.

Likewise, some leading scientists are leaving their academic appointments to join the new private enterprises in plant biotechnology.

These events are happening because the private sector wants to use biotechnology to accelerate its growth in agri-business in the short-term. Nonetheless, funds to support basic and strategic research by public researchers are needed for a long-term sustainable transfer of public goods to the private sector or other users.

BIOINFORMATICS

Another important factor in the successes of the genetic improvement of crops was the development of fast and more reliable computers, which allowed easier management and analysis of data as well as publication of scientific reports. The impact of the informatic revolution in crop improvement can be partially assessed by counting the number of publications indexed in Plant Breeding Abstracts. There was ca. 22-fold increase of publications in the 1930-1997 period. It was in the 1970s that indexed publications in plant breeding exceeded 10,000 per year. More publications and easy means for retrieving this information accounted for such growth of knowledge dissemination in plant genetics and breeding.

Today, rapid information exchange has been facilitated with electronic mail and access to the internet to read electronic publications such as this journal. Nowadays, information technology and DNA science are beginning to fuse into a single operation. Computers are deciphering, and organizing the huge genetic information that may become "the raw resource of the emerging biotech economy" in the next century. Scientists working in the new field of "bioinformatics" are developing biological data banks to download the genetic information accumulated during millions of years of life evolution, and perhaps reconstruct some of the living organisms of the natural world.

PLANT GENOMICS

Plant Genomics refers to the investigations of whole genomes by integrating genetics with informatics and automated systems. This term defined by the development of biotechnology. Genomic research aims to elucidate the structure, function and evolution of past and present genomes. Some of the most dynamic fields concerning agriculture are the sequencing of plant genomes, comparative mapping across species with genetic markers, and objective assisted breeding after identifying candidate genes or chromosome regions for further manipulations.

As a result of genomics, the concept of gene pools has been enlarged to include transgenes and native exotic gene pools that are becoming available through comparative analysis of plant biological repertoires. Understanding the biological traits of one species may enhance the ability to achieve high productivity or better product quality in another organism. DNA markers and gene sequencing provides quantitative means to determine the extent of genetic diversity and to establish objective phylogenetic relationships among organisms. 'Gene chips' and transposon tagging will provide new dimensions for investigating gene expression. Molecular biologists will study not only individual genes but how circuits of interacting genes in different pathways control the spectrum of genetic diversity in any crop species.

FARMING

The aims of applied plant science research for agriculture are to enhance crop yields, improve food quality, and preserve the environment where human beings and other organisms live. The best way for conservation of plant biodiversity and its environment, would be to achieve high crop productivity per unit area. In this regard, Briggs reported that as yields treble, soil erosion per ton of food decreases by two-thirds. There has been a significant yield improvement owing to enhanced crop husbandry, but in the next years progress will be achieved by changing plants that could be more suitable to sustainable and environmentally friendly farming systems.

Agro-chemical corporations are developing pest and disease resistant transgenic crops to avoid pollution with pesticides in the farming system. Furthermore, food quality will become more important than crop productivity in a wealthy society. Consumers will prefer transgenic crops if they have the desired characteristics. Genetic improvement through biotechnology needs conventional breeding because

1) the elite cultivars will be the parents of the next generation of improved genotypes,
2) field testing across locations or cropping systems and over years will be needed to determine the best selections due to the genotype-byenvironment interaction.

Transgenes must be viewed as improvements rather than replacements for elite germplasm. Indeed, genetic engineering may provide a means to add value by introducing synthetic or natural genes that enhance crop quality and yield, as well as protect the plant against pest and diseases. Farmers will pay more for transgenic crop propagules if they obtain extra-income after adopting biotech-derived products. For example, seeds of insect resistant transgenic crops will be more expensive than those of available cultivars but the farmer will not need to apply pesticides in their transgenic fields. Of course, patents make transgenic seeds more expensive but also farmer's benefits may be higher.

DNA BANKING, GENE BANKS AND VIRTUAL PLANT BREEDING

The advances in gene isolation and sequencing in many plant species allows to envisage that within a few years, gene-bank curators may replace their large cold stores of seeds with crop DNA sequences that will be electronically stored. The characterization of plant genomes will ultimately create a true gene bank, which should possess a large and accessible gene inventory of today's non-characterized crop gene pools. Of course, seed banks of comprehensively investigated stocks should remain because geneticists and plant breeders, the main users of gene banks, will need this germplasm for their work. Genomics may accelerate the utilization of candidate genes available at these gene banks through transformation without barriers across plant species or other living kingdoms.

Nonetheless, genetic engineering should be seen as one of the methods of plant breeding that permits the direct alteration and re-building of a crop population. "Shutting-off" genes coding for undesired characteristics may be another application of transgenics in crop improvement. Plant breeders will change their modus operandi with the development of objective marker-assisted introgression and selection methods. Backcross breeding will be shortened by eliminating undesired chromosome segments of the donor parent or selecting for more chromosome regions of the recurrent parent.

Parents of elite crosses may be chosen based on a combination of DNA markers and phenotypic assessment in a selection index, such as best linear unbiased predictors. To achieve success in these endeavours, cheap, easy, decentralized, and rapid diagnostic marker procedures are required. There are many areas of basic and strategic research in plant breeding and genetics that are being facilitated by marker-aided analysis. With molecular markers, plant biologists are reviewing crop evolution and gathering new knowledge. Such information should be incorporated into genetic enhancement programmes, especially those with an evolutionary breeding scheme.

Likewise, plant ideotypes for each crop should drive the work of plant breeders. Specific plant morphotypes have been defined in rice and wheat based on accumulated knowledge of crop physiology and crop protection. The needed characteristics required to develop improved plant prototypes ensuing from such a 'virtual breeding' approach may be available in gene banks of the crop or in those of other species. Otherwise, breeders may obtain novel transgenes to develop the required ideotype. Nowadays, the finding of new genes that add value to agricultural products seems to be very important in the private agri-business. Unique gene databases are being assembled by the industry with the massive amount of data generated by genomics research.

A new term 'biosource' was coined recently to refer to a fast and effective licensed technology of pinpointing genes. With this method, a 'benign' virus infects a plant with a specific gene that allows researchers to observe directly its phenotype. Biosource replaces the standard time-consuming approach of first mapping a gene to subsequently determine its exact function. Gene identification in DNA libraries coupled with biosource technology and an enhanced ability to put genes into plants will be routine for improving crops in the next decade. Genomics may provide a means for the elucidation of important functions that are essential for crop adaptedness.

Regions of the world should be mapped by combining data of geographical information systems, crop performance, and genome characterization in each environment. In this way, plant breeders can develop new cultivars with the appropriate genes that improve fitness of the promising selections. Fine-tuning plant responses to distinct environments may enhance crop productivity. Development of cultivars with a wide range of adaptation will allow farming in marginal lands. Likewise, research advances in gene regulation, especially those processses concerning plant development patterns, will help breeders to fit genotypes in specific environments.

Photoperiod insensitivity, flowering initiation, vernalization, cold acclimation, heat tolerance, host response to parasites and predators, are some of the characteristics in which advanced knowledge may be acquired by combining molecular biology, plant physiology and anatomy, crop protection, and genomics.

Multidisciplinary co-operation among researchers will provide the required holistic approach to facilitate research progress in these subjects.

PHARMING AND FARMER-CEUTICALS

Growth of cities in the developed world has already replaced farmland with shopping malls, parking lots, and housing developments. Peri-urban agriculture and home gardening are becoming very important for national food security in the developing world as a result of rapid urban expansion. Hence, new cultivars will be needed to fit into intensive production systems, which may provide the food required to satisfy urban world demands of the next century.

Specific plant architecture, tolerance to urban pollution, efficient nutrient uptake, and crop acclimatization to new substrates for growing are, among others, the plant characteristics required for this kind of agriculture. Genes controlling these characteristics may be available in gene banks for further cross breeding, which can be assisted by genomics. Peri-urban and home garden "farmers" will have to adapt to new demands from emerging urban populations with higher income. These consumers may request a more varied diet.

For example, food crops with low fats, and high in specific amino acids may be needed to satisfy people who wish to change their eating habits. If genes controlling these characteristics do not exist in a specific crop pool they may be incorporated into the breeding pool using transgenics. Some publications anticipated that in the next millennium food will not need to be harvested from farmer's fields. Tissue culture of certain parts of the plant may provide a means to achieve success in this endeavour. For example, edible portions of fruit crops could be grown in vitro. A steady and cheap supply of these edible plant parts will be required in this new agri-business. It will take some time before such a process can be scaledup for commercial output.

Often plants provide the raw materials for agro-industry, and not only for food or fibre processing. Active ingredients of plants have been transformed into commercial products such as medicines, solvents, dyes, and non-cooking oils for many years. Hence, it would not be surprising to see, in few years from now, entire farms without food crops but growing transgenic plants to produce new products, e.g. edible plastic from peas or plant oils to manufacture hydraulic fluids and nylon. This new rural activity may result in important changes in the national economic sector.

'Pharming' has been added to the dictionary to indicate a new kind of system to obtain medicines. For example, oral vaccines appear to be a convenient delivery system for vaccination throughout the world. Biotechnology has been used to engineer plants that contain a gene derived from a human pathogen. An antigenic protein encoded by this foreign DNA can accumulate in the resultant plant tissues. Results from preclinical trials showed that antigenic proteins harvested from transgenic plants were able to keep the immunogenic properties if purified. These antigenic proteins caused the production of specific antibodies in injected mice. Mice, which ate these transgenic plant tissues, also showed also a mucosal immune response.

Ability of transgenic food crops to induce protective immunity in mice against a bacterial enterotoxin such as cholera toxin B subunit pentamer with affinity for GMI-ganglioside. Also, potato tubers have been used successfully as a biofactory for high-level output of a recombinant single chain antibody.

TRANSGENIC CROPS RISK ASSESSMENT

The issue of releasing genetically modified plants (GMP) into the farming system has become particularly agitated by lobbyist groups in Europe despite widespread cultivation of such crops in North America and elsewhere. Scientists must realise that the general public are concerned that an uncautious approach to the manipulation and cultivation of transgenic crops may affect biodiversity and its sustainable utilization in the farming system, e.g. loss of variability and viability. People also want that their views about applications of biotechnology for improving agriculture are listened irrespective of their knowledge in the subject.

Moreover, farmers are afraid that negative propaganda jeopardizes the public image of their products. Scientists and policy markers should not forget that people's acceptability is the most important component of the general public assessment of risk, which includes both uncertainty and negative consequences. This acceptability depends on cultural factors because people's views change according to time and location. The process of risk assessment in agro-chemical consists of

— hazard identification,
— exposure assessment,
— effect's management,
— risk characterization, and
— risk management.

However, transgenic crops may be able to invade and multiply in many habitats. Hence, this risk assessment of a genetically modified living organism (GMLO) must consider other characteristics not included when assessing the release of non-living compounds to the environment, e.g. horizontal gene transfer between transgenic crops and wild related species. Scientific risk assessment of transgenic crops must be strictly performed and precautionary principles should be considered in the decision making process. In the industrialized world, this precautionary principle is a key component of the response to the unforeseen human and environmental impact, which may occur by introducing into the system new advances ensuing from research and technology development.

In Norway, an unique legislation advocates that "the production and use of GMO should be ethically and socially justifiable in accordance with the principle of sustainable development" as well as "safe to humans and to the environment". By applying this framework, marketing applications of GMO could be rejected if insufficient documentation regarding ecological and heath aspects was submitted by the producer. What are the potential ecological risks associated with the release of GMP into the farming system? These are of course a very large number of potential risks, However, perhaps the two most important risks are:

1) GMP establishes in semi- or natural habitats, and
2) inserted transgenes incorporate into other species, thereby affecting non-target organisms in farms or natural habitats.

Hierarchical test protocols have been proposed to assess the risks of releasing GMP. Such protocols require knowledge about evolutionary history, morphology, life-history characteristics, pollination or breeding system, genetransfer likelihood, natural hybridization, recruitment and vegetative propagation of a chosen species. Likewise, producers should provide, to facilitate this risk assessment, additional information regarding biochemical, physiological, and morphological changes owing to inserted genes, along with a list and description of marker and reporter genes included in the transgenic plant.

It would also be important to add details concerning when and in which plant tissues or organs will be expressed the modified function or phenotype. Nonetheless, people must also know that scientists assessing risks of transgenic crops may extrapolate the outcome or results from simple shorttime experiments into complex long-term natural- or farming systems. Investigations about gene flow and competing ability of transgenic crops may be easily addressed through short-term experiments. However, the assessment of the environmental impact of GMP requires a long-term, expensive, holistic research. Computer modelling, which integrates knowledge about gene flow, competing ability, spread of transgenes to weedy species, and cultural practices in the farming system, may provide an alternative means for long-term risk assessment of releasing GMP into the environment.

Consumer concern about transgenic crops also focuses on their safety as food, especially if modifications could influence their metabolism or health. In this regard, transgenic plants without selectable markers, such as antibiotic resistance genes, are needed to convince GMPsceptics of the advantages of genetic engineering for crop improvement. In this way, their criticism concerning the potential risks of transgenic crops could be overcome.

BIBLIOGRAPHY

Eujayl, I. & Baum, M. & Erskine W & Pehu, E & Muehlbauer F.J. *The use of RAPD markesrs for lentil genetic mapping and the evaluaionof distored F2 segregation*, Euphytica 96 (1997) 405-412.

Evers, Aino-Maija & Tuuri, H. & Hägg, M. & Plaami, S. & Häkkinen, U. & Talvitie, H. "Soil forming and plant density effects on carrot yield and internal quality", *Plant foods for human nutrition*. - Dordrecht : Martinus Nijhoff Publishers. 51 (1997) s. 283-294

Kiviharju, Elina & Puolimatka, Matti & Pehu, Eija, "Regeneration of another-derived plants of Avena sterilis". *Plant cell, tissue and organ culture*. - Dordrecht ISSN 0167-6857. 48 (1997), s. 147-152

Kleemola, J., Teittinen, M. & Karvonen, T. "Modelling crop growth and biomass parti-tioning to shoots and roots in relation to nitrogen and water availability, using a maximization principle" *I. Model description and validation. Plant and Soil* (1996) 185:101-111.

Koivu, Kimmo & Valkonen, Jari & Suomaa, S. & Tavazza, R. & Pehu, Eija

Rokka, V-M., Clark, M.S., Knudson, D.L., Pehu, E. & Lapitan, N. L.V. *Cytological and molecular characterization of repetitive DNA sequences of Solanum brevidens and Solanum Tuberosum.* Genome 41 (1998) 1-8.

Seppänen, Mervi & Majaharju, Minna & Somersalo, Susanne & Pehu, Eija "Freezing tolerance, cold acclimation and oxidative stress in potato: paraquat tolerance is related to acclimation but is a poor indicator of freezing tolerance" *Physiologia plantarum.* - Copenhagen ISSN 0031-9317.102: 1998 s. 454-460

Valkonen, Jari & Keskitalo, Marjo & Vasara, Tuija & Pietilä, Leena "Potato glycoalkaloids: a burden or a blessing?". *Critical reviews in plant sciences*. - Boca Raton, FL : CRC Press ISSN 0735-2689. 15 (1996): 1 s. 1-20

Watanabe, Kazuo N. & Pehu, Eija, *Plant biotechnology and plant genetic resources for sustainability and productivity* / Kazuo N. Watanabe, Eija Pehu. - New York : Academic Press, 1997. - 247 s. - ISBN 0-12-737145-1

________________. & Orrillo, M. & Vega, S. & Valkonen, Jari & Pehu, Eija & Hurtado, A. & Tanksley, S. D, "Overcoming crossing barriers between nontuber-bearing and tuber-bearing Solanum species: towards potato germplasm enhancement with a broad spectrum of solanaceous genetic resources" *Genome*. -Ottawa ISSN 0831-2796. 38 (1995) s. 27-35

INDEX